GAZING UPON SHEBA'S BREASTS

Exploring Our African Origins

by
David Price Williams

Gazing Upon Sheba's Breasts: Exploring Our African Origins © 2016 David Price Williams & Markosia Enterprises, Ltd. All Rights Reserved. Reproduction of any part of this work by any means without the written permission of the publisher is expressly forbidden. Published by Markosia Enterprises, PO BOX 3477, Barnet, Hertfordshire, EN5 9HN. SECOND PRINTING, February 2017.
Harry Markos, Director.

Paperback: ISBN 978-1-909276-71-0
Hardback: ISBN 978-1-909276-70-3
eBook: ISBN 978-1-909276-87-1

Book design by: Ian Sharman

FRONT COVER PICTURE:
The Trance Dance: four men dance whilst two shamans pass into an altered state of consciousness. Nsangweni Rock Shelter, Komati Valley, North West Swaziland. (Picture by DPW).

BACK COVER PICTURE:
DPW at Nkhaba.

www.markosia.com

Second Edition

DEDICATION

For Sue, who gave me the gift, and for Alice and David, who came to share it with me.

CONTENTS

Out of Africa	i
Pronunciation	iii
Chapter 1 States of Altered Consciousness	1
Chapter 2 Images of Power	9
Chapter 3 The Place of Winds	24
Chapter 4 Christmas Hand-axes	27
Chapter 5 Lion of the Nation	47
Chapter 6 Making a Plan	57
Chapter 7 Dr Watson's Patent Colluvium	70
Chapter 8 Unpronounceable Scholarship	82
Chapter 9 Talking to the Trees	93
Chapter 10 Incised Meanders	107
Chapter 11 Up a Hill with Sieves	118
Chapter 12 A Whole Lot of Scraping Going On	129
Chapter 13 Mlawula Dreams	139
Chapter 14 Rhinos up a Tree	152
Chapter 15 The Pipless Lemon Squeezy	160
Chapter 16 General Weakness	170
Chapter 17 Getting to the Bottom of it All	180
Chapter 18 Royal Geographical Society	185
Chapter 19 Evolution and the Methodists	200
Chapter 20 Beginnings and Endings	210
Afterword	220
Map of Swaziland	221
The Old Farm with 'Stoep' & the Long House	222
Insangweni Rock Paintings, Toko & Watty	223
Swaziland National Museum	224
Betsy Woodwell's oil collage of DPW at Mlawula	225

OUT OF AFRICA

Every year we read about the discovery of a new specimen of early human found in Africa, a new skull perhaps or a strand of DNA, which is helping us to piece together the story of where we originated. Apparently, as scientists now think, we came from Africa, all of us; beginning four or five million years ago it was in Africa that we emerged onto the world stage. And with the later development of stone tool technology we were eventually able to migrate from Africa and colonize the rest of the globe.

However, one hundred years ago that wasn't what people believed at all; Africa was not even remotely considered a possible cradle of humankind. That location, we were told in all seriousness, was actually in England, in East Sussex to be precise, where fossils had been found in a gravel pit at Piltdown near Uckfield. 'Piltdown Man', as it became known, was 'discovered' in 1912 by a local solicitor, Charles Dawson, and was widely acclaimed by the top medical practitioners of the day to be 'the missing link', the half-way phase between the apes and modern humans, although contemporary humorous cartoons often depicted this honour being bestowed upon opposing politicians at Westminster!

But in November 1924 something was stirring in South Africa. The new professor of anatomy at the University of the Witwatersrand in Johannesburg, Raymond Dart, was handed a cardboard box containing fossil baboon skulls which had been found at a lime-works at Taung, not far from Kimberley in the Orange Free State. On top of the box was the lime-hardened cast of a small brain as well as a block of travertine from which, when cleaned, emerged the face of a young child. Dart recognised them to belong to a very early form of human and from this he claimed that Africa, not Britain, was the origin of our lineage. Dart named his discovery *Australopithecus africanus*, the Southern Ape of Africa. That was the beginning.

But it wasn't until several decades later, after a number of adult Australopithecine specimens had been unearthed from other South African lime-works sites, that Dart's original idea was taken seriously and the concept of our African genesis began to become fashionable. As a result, the Piltdown specimen was re-examined at the Natural History Museum in London and in 1953 it was announced that 'Piltdown Man' was a forgery. It turned out to be the biggest hoax ever perpetrated upon the scientific community and so misled popular opinion that it probably held back the proper search for the African emergence of human-kind for half a century.

More recently, many new fragments of evidence have been recovered, both in southern and eastern Africa, so that the whole drama of our introduction into the global theatre has now been carefully mapped out. While debates still rage about the precise order of events, Africa has been firmly established as our birthplace.

And what about the artefactual discoveries? In the nineteen-sixties and seventies new radiometric dating techniques started to demonstrate that major advances in the manufacture of stone tools also began on the African continent. These were the vital technological improvements our ancestors made to their very existence, to the way they made a living in this vast untamed arena, and the way they competed among the elephants and rhinos, leopards and lions and all the rest of the terrifying bestiary which characterized their world.

What impelled them to make these changes? Was their world the same as ours today, or was it very different? Why did human evolution take place at all, physically and, even more significantly, culturally, on the continent of Africa? This memoir is about that very question. It's the story of a major research project carried out over a period of more than a decade and based in the beautiful Kingdom of Swaziland, home of the AmaSwazi. Their own traditional way of life is focused on the Royal Residence at Lobamba, overlooked by the twin peaks known locally, after Rider Haggard's description in 'King Solomon's Mines', as 'Sheba's Breasts', evoking literary memories of his illustrious visitor, the Queen of Sheba.

Perhaps everyone who visits wild Africa begins to experience a strange sense of recognition, a feeling of belonging in this timeless landscape. It happened to the author forty years ago. It is an overwhelming sensation of coming home, which in a way is only natural. For Africa is where we all began!

PRONUNCIATION

SiSwati is one of the five Nguni languages of south-east Africa, as is siZulu, and they in turn are part of the wider Bantu language group. During protracted exposure to the Khoi/San (Hottentot/Bushman) peoples, these Nguni speakers picked up a small number of the clicks which strongly characterized the indigenous tribal languages. The only one reproduced here is the strong alveolar click, the sound of a popping cork, represented by the symbol '!'

SiSwati and siZulu also make use of an aspirated 'L' sound formed with the side of the tongue, as in the Welsh 'Llangollen', represented in that case by the double 'Ll'. In the African equivalents the sound is written 'Hl' as in Hlatikulu. Generally, Europeans are unable to pronounce this sound and instead use the form 'Sh' as in Shlatikulu.

Plurals in siSwati are formed with either the prefix *ba-* as in *bafundzisi*, 'instructors', the prefix *ema-* as in *emajaha* 'young men' or the prefix tin- as in *tindlovu*, 'elephants'. Dialectically, in siZulu this last plural would be rendered *zindlovu*.

The siSwati orthography follows David Rycroft's 'Concise SiSwati Dictionary' (J.L. van Schaik 1981), except the word for 'father', which Rycroft renders *babé* with the diacritic, but which would in normal text read 'babe'. Since 'Hey babe' would lead to endless misunderstandings, I wrote it as bahbe, the long 'ah' better reproducing the sound used. Also, I have used the Swazi spelling for the 'Lubombo' Mountains.

The main *ncwala* anthem title is taken from Hilda Kuper's book 'An African Aristocracy', Oxford 1947.

For spelling and other details from Zululand I am grateful to the late David Rattray and to the Anglo Zulu War Historical Society.

CHAPTER ONE

STATES OF ALTERED CONSCIOUSNESS

Suddenly in the midst of spiritual darkness and oppression, there seem at moments a flash of light in the brain, and with extraordinary impetus all the vital forces suddenly begin working at their highest tension. The sense of life and the consciousness of self are multiplied ten times at these moments which passed like a flash of lightning.

Fyodor Dostoevsky : The Idiot

It was to be my greatest gift, and I began by hating it. Africa, I mean. Initially, at the start of my first visit, I loathed it. Looking back now, across the intervening forty years or so, it is difficult to say why. But then, in a different century, a different millennium even, Africa's ancient landscapes seemed so strange to me, so far away in time and place. To me, in every respect, Africa was unfamiliar, undefined, incomprehensible - even hostile. Take the appearance of the land to begin with, the tilt and rake of the mountains, the lime green of the maize stalks, the vermilion earths - that startling chromatic contrast of verdant foliage springing from ancient granite. For me it was unrecognisable, beyond the knowable.

And yet I was to become familiar with Africa as though it were my own skin. I was bitten by Africa one tempestuous night that first December, and it got into my blood, like malaria. I fought with Africa and we came to understand each other. Slowly, inexorably, she revealed her secrets. I learnt her ways, profound and ageless. Gradually I began to unlock her deeper mysteries, and, with that, I stumbled upon mankind's own furthest history. We came from Africa, all of us. Africa has made us and shaped us. This tropical savannah is the land of our own collective beginning. Taking back-bearings, I see now that it was only natural that in the end Africa was where I too would find myself, physically, spiritually and metaphorically. But that was still in the future.

* * * *

One summer in the mid-seventies I had married. My wife Sue had been brought up on a farm in Swaziland, a tiny country in southeast Africa I had never heard of, bordering the Republic of South Africa to the west and Mozambique to the east. I looked it up in an old school atlas. It was barely visible, an insignificant circle the size of Yorkshire or Delaware in a vast continent more than three times the surface area of the United States. Even the name of its capital, 'Mbabane,' printed large enough to be legible, obliterated the shape of its country's outline. That's how small it was. That's where we were to go for our first nuptial Christmas, to meet the 'in-laws' who had never encountered an archaeologist before, let alone one married to their much loved daughter.

"How large is the farm?" I had asked casually, on our first evening date. I came from the borders of Shropshire, from a well-heeled farming community, though I had personally been brought up in circumstances best described as cultured penury. But at least with talk of farms I felt we had something vaguely in common.

"Three thousand acres," came her quiet reply.

"THREE THOUSAND ACRES!" I spluttered. Tiny slivers of kebab peppered the table cloth. "Three thousand acres?"

In the mist-haunted upper Severn valley where I came from I had seen people live like Croesus of old with barely a hundred acres. But three thousand acres?

"Oh, pay no attention to that," she had said, waving her hand dismissively. "Daddy owes it all to the bank. Mummy and Daddy live in a mud house with a tin roof. The farm is in hock to all sorts of people."

Affecting an unusual nonchalance, I continued;"Of course they do! Of course it is!"

But she had been right about the farm. When we arrived in the Malkerns Valley in Swaziland that same December, so many of the farmers in the region were deeply in hock to Swaziland Milling Company, Swaziland Agricultural Supplies, Barclays Bank of Swaziland, Swaziland Warehouse and sundry other Swaziland chandlers, seed merchants and loan sharks from whom they had all borrowed to keep body and soul together in the ultra-high-risk business of farming in the tropics. Every pestilence known to man – red-billed queleas, helmeted guinea fowl, army worm, locusts, hailstones the size of tennis balls, fungal spores, root rot, leaf blight drought, torrential rain, parching sun, fire, wind, lightening like Zeus' thunderbolts – had systematically, over the years, nibbled, eaten, withered, eroded, battered, scorched and generally lived on, in and under every crop they had planted.

It was like a perpetual replay of the ten plagues of Egypt. And now the farms were all in pawn. And the 'in-laws' did indeed live in a house of clay and wattle made, with a dented, red-oxide, corrugated iron roof. Yet it was there, on the simple veranda of that dilapidated homestead in the Malkerns Valley that my African conversion was to take place - my light in the sky, literally.

* * * *

We had arrived in Johannesburg that Christmas after a long night-flight from London, putting down *en route* in the fetid heat of the Cape Verde Islands, a barren, volcanic archipelago off the coast of Senegal. At two in the morning, we had deplaned to refuel at Ilha do Sal and had made our way through the stifling murk to a large, ill-lit hutment with a scatter of rickety plastic chairs. Another group had already occupied all the available seating. They turned out to be one hundred and sixty young Cubans flying to Russia in a decrepit Ilyushin jet for indoctrination in the liberating ways of the Supreme Soviet. We represented ideologies that passed in the night. It was a dismal encounter. We looked at each other with a weary incomprehension before filing back on to our Boeing and lifting off into the moonlit sky.

I met the 'in-laws' waiting at the airport, who to their credit made no immediate comment on the hirsute archaeologist of dubious prospects whom they had just gained as a son-in-law. Gin-and-tonics at ten o'clock in the morning seemed to be a propitious beginning, but then we had to endure the long and gruelling drive across the monotonous flatness of what was then the central Transvaal, low ridge after low ridge of open road, flanked by straggling pink and white cosmos and small stands of jaded wattle trees. Apart from isolated mine dumps and endless maize fields, the lightly undulating landscape wore a soiled appearance, empty and featureless. Rarely, in the middle of this deserted scene, there was a cluster of two or three African mud hovels with flat, rusted tin roofs held down with stones and old car tires. Far off, hidden from view behind clumps of blue gums was the occasional 'white' farmhouse. Here and there a wind pump turned languorously in the light air. It was two hundred miles of relentless tedium, punctuated only by far-spaced Afrikaner 'dorps' (villages) like Breyten and Carolina. If this was Africa, to me it totally lacked animation. It had been stripped of anything recognisably exotic, or tropical. It was all so desperately unimaginative and depressing.

Farm notices alleviated the dreariness, with improbable and unfamiliar European names like Allesforloren (All is lost), Helpmekaar (Help each other), Bultpan (Hump pond), Vogelstruisfontein (Ostrich Springs), and

Van Wyksvlei (Van Wyk's bog). Then gradually after some hours the scenery changed. As we neared the Swaziland border rocky knolls began to appear, and the farm names changed too - Lochiel, Waverley and Glenmore. The bulk of Ngwenya Mountain reared up in front of us, a vast iron massif which turned the earth a deep metallic crimson. I was about to experience the 'real Africa' for the first time. We crossed the no-man's land between the dorp-inspired Afrikaner border post of the Suid-Afrikaanse Republiek to the simple bureau of the Kingdom of Swaziland Immigration and Customs Post.

I was later to learn that Swaziland had never been part of South Africa. In 1903, after the ending of the Boer war, it had been wrenched from the unruly grip of a rag-tag bunch of Anglo-Boer speculators and farmers and had notionally been given the status of a British Crown Protectorate, though no formal instrument of annexation or protection seems ever to have been signed. Sixty five years later, in 1968, it had achieved its full independence. There were elderly Swazi councillors who could remember the British Empire arriving and departing. By the time I got there, Swaziland had reverted to being a traditional African kingdom.

Inside, the immigration hut was a milling and unrestrained chaos; non-segregated, it was a heaving mass of multi-cultural humanity. There were rotund *bafati* (married women), busy with bulging carrier bags and parcels, tiny curly-headed infants tied to their backs with colourful tartan shawls. There were truck-drivers aimlessly waving carnets and customs papers amid the hubbub. Backpackers jostled with African students going home for Christmas. Swazi warriors in full regalia with sharp-pointed assegais mingled with golf-playing white supremacists trying to ignore the din and throng around them. And there was us.

We finally approached the counter, which looked like a cross between a raucous public bar and Brixton Police Station. The smartly-dressed Swazi officers smiled encouragingly at the assembled press of people, flashing their ivory teeth genially at all comers. Passports were stamped, several times. The thud, thud, thud of a rubber stamp on an ink pad, then thumped on the paper, is one of the great legacies the British Empire has bestowed upon its ex-colonies. A government official in Africa without a rubber stamp is as impotent as a point-duty policeman with no arms. Worming our way back through the seething throng we arrived in Swaziland.

* * * *

The real Africa opened up before me. The names on the roadside proclaimed a foreign, non-European land – Motshane, Sidwashini, Mahlanya. We drove through Mbabane, the mountain capital surrounded by mighty granite tors - boulder-strewn koppies placed as though they had fallen from a giant's apron. At the top of the main, if not the only, street we passed Mbabane's premier hotel, 'The Tavern'. Incongruously, it was tricked out in mock half-timbered Tudor and for all I knew it had been erected with prefabricated sections shipped from some industrial estate in Kettering or Kidderminster. But just down the street the pavement was a sea of African faces. Some of the men were wearing skirts with animal-skin sporrans, tweed jackets and brogues. The scene bore no resemblance to anything I had encountered before. I was completely out of my element.

Shortly after leaving the tiny one-traffic-light metropolis, we came to the precipitous edge of the great African Escarpment, the Malagwane Hill, and peered down into the Ezulwini Valley, the Place of Heaven. I came to love that view, but at this first sight it was like staring into the Abyss. The ground fell away almost two thousand feet into a colossal scuttle-shaped basin, lined on both sides by glowering crags. On the left stood the mountain range of Mdzimba, huge peels of grey-black exfoliated granite effulgent with streams like sheets of quicksilver. On the right, at the end of a long ridge, and as though directly out of *'King Solomon's Mines,'* rose the inscrutable twin peaks of Lugogo, 'the Grandmother', known colloquially as 'Sheba's Breasts'. Rider Haggard surely must have been describing these phenomena when in 1885 he wrote:

> 'For there, glittering like silver in the morning sun, were Sheba's Breasts. These mountains standing thus, like the pillars of a gigantic gateway, are shaped exactly like a woman's breasts.'

And indeed they are. The whole scene was an alternating scooped out and precipitous beetling mass of vivid emerald green. Along the foothills, as far as the eye could see, there were Swazi kraals, clusters of hemispherical beehive huts of dull, grey thatch. Each had a cattle byre nearby, made from boughs and tree trunks, shaded by the spreading branches of African flamboyants with their vivid electric-orange flowers, bright flame against the green luxuriance of the high mealie plants (maize). In the far distance rose the curved spike of a mountain they call Luphondvo, the Rhino's Horn, which overlooks the Malkerns valley, our destination. This was where the three thousand acres of African promise lay, my wife's family farm among the guava trees, pineapple plants and mealie fields. It was all fashioned in a mystifying melange of outlandish names, strange rocks, a dazzling array of colours and cultures, all totally unfamiliar to me. Of this landscape and this people I could not read one word. Comprehension failed me.

Next day rain poured from the sky like water from a bucket and splashed shrilly from the down-spouts into spreading pools across the bare, dark-red earth. The clouds were low, leaden and threatening. They obscured the mountain tops and sat brooding and drifting over the Malkerns Valley. All around, the farm roads were churned up into a livid maroon mass, contrasting with the now-muted green of the crops. The heat and humidity acted as their own sodden dampener on this tropical yuletide. It was the very antithesis of what I understood to be Christmas - the Christmas card emblems of snow drifts against mullioned window panes, tartan-rugged stage-coach travellers hallooing through white flurries, carollers in chequered mufflers carrying lanterns, braving the seasonal, crisp, wintry weather. I don't know who had dreamed up all this arcane iconography of Yule logs and robins, but it had been engrained in me since I was nought. The real Christmas and the cold weather were inextricably linked in my mind. But frosty air was the last thing they ever breathed in Malkerns. It was all so desperately unfamiliar, and so disheartening.

The storm brooded the whole day. That evening it worked itself up to a tempestuous crescendo, with thunder and lightning reminiscent of the well-known British Army film clip from the opening salvos at the Battle of El Alamein. The downpour would have done justice to a Cecil B. de Mille remake of Noah's Flood. It rattled on the tin roof like so many demented demons with ball peen hammers such that normal conversation was rendered impossible for the din. We sat in the drawing room in antediluvian, chintz arm-chairs, apparently meant to believe that this was all in the normal course of events. Thunderclaps cracked the air, and there was the disquieting crash and hiss of lightning strikes in the near vicinity. The smell of high- voltage ozone filled the room and the power supply kept cutting out, plunging us into periodic darkness. It was terrifying, like being present at the Creation. I tucked into my gin and tonic with growing alarm. I decided right there that I hated it all.

But then, just as in Beethoven's Pastoral Symphony, there came the calm after the storm. At first in the abrupt quiet and stillness it seemed that we had all been smitten with collective deafness. I walked out onto the ageing stoep, a small red-polished veranda opening onto the garden and fields beyond. Sheet lightening still flashed incessantly but now silently among distant clouds, the flickering light leaping among the dark masses like a far-away tiered firework display. As I remained, alone and watching, I suddenly became aware of a miraculous fragrance. The warm, dank air reeked of growing, of the mulch and must of vegetation. There was a thick scent of green, of budding and new leaves, of perfumed frangipane and rose petals, of the dripping earth returning to life out there in the darkness. The storm had awakened all the rising sap and the plant life was quaffing the liquid of nature. The clouds above parted and a myriad tiny points of white fire emerged from the black, fleeing shadows. A vast canopy of the brightest, cleanest, most numerous stars I had ever seen pierced the firmament. I stood transfixed.

At that moment my hearing seemed to return, as though someone had thrown a switch or lifted a maestro's baton, and at once a truly astonishing barrage of noise reached my ears. A slow, deep, throbbing clamour pulsated in the air. What sounded like a million toads in the irrigation furrow croaked into the night sky, *Bufo gutteralis* at their rattling, choral best, accompanied by the black-smith pink, pink, pink of innumerable tiny reed frogs. A buck coughed in the blackness, and the last drops of rain tinkled in the pools. Palpable waves of sound washed over me. My senses - my smell, sight and hearing – were all so intensely assailed I became entranced. I called my wife onto the veranda. 'Look! Listen!' I whispered, 'Africa!' I was totally mesmerised. She had just given me my greatest Christmas gift. Inexplicably, I suddenly felt I had come home.

* * * *

It is strange to consider, looking back, that in that moment I had recognized something so elemental, despite having never actually experienced it before. I had immediately grasped how significant it was without knowing anything at all of its content. Today, as I walk through the African bush, able to identify trees by their scientific name, their common name, their African name, the soil type, the rainfall, the temperature range and the altitude; when I can thrill to every rock structure, give its age, foundation, geological importance, name the birds and their migrating habits, pick up a two million year old stone tool and know that this is an implement from the dawn of humanity, it is difficult to remember a time of not knowing. But in that first mesmeric, toad-croaking, star-lit darkness of unmitigated ignorance, a voracious appetite was born within me, an acute hunger to comprehend what this vast continent was all about. It was like suddenly and desperately needing to communicate with people who converse in a language you have never heard before but with whom you must speak; where from being totally fluent in your own tongue you are reduced to a mere inarticulate show of signs and mute gestures. The lack of knowledge was absolute; the desire to know frantic.

The result of my Damascene conversion was that from passively accepting my new circumstance I became immediately and furiously active in finding people who could describe this Africa to me and cure me of my recently acquired intellectual blindness. I was passed from one Ananias to another by enthusiastic, newly-discovered, if somewhat perplexed relatives. And so it was that I heard about the Bushman paintings of the Komati Valley. I first became aware of them like some distant but arousing bat-squeak from the remote past, a stirring rustle from a far-off people in a time beyond history. It was a barely audible echo, a faint cry from an earlier kin overheard indistinctly as through a dense mist.

CHAPTER TWO

IMAGES OF POWER

The fact that Bushman paintings illustrate Bushman mythology gives to Bushman art a higher character and teaches us to look upon its products not as the mere daubing of figures for idle pastime but as an attempt at a truly artistic conception of the ideas which most deeply filled the Bushman mind with religious feelings.

Wilhelm Bleek 1874

Hearing about the Bushman paintings, I at once set off in odyssean pursuit, as though drawn by some seductive siren. Well, I say odyssean pursuit, but the manner of this odyssey was to be rather less than romantically classical. Where I was intending to go was considered to be 'in the wilderness'. What I would need, I was informed, was a Land Rover. In the muddy yard at the back of the house, amid the rusting farm machinery, stood the barque of my dreams, or was it nightmares – a dilapidated Series Two Land Rover station wagon painted a dull blue. It had once been a police Land Rover that in its long-passed, law-enforcing heyday had been driven at full throttle over the most unforgiving landscape on earth until even the local constabulary had finally admitted that as a conveyance it had reached its last bend in the road. My relatives had bought it on that last bend to live out its final days on the farm, in a way being put out to grass like some ageing stallion in its dotage, which having completed his allotted stud can spend the rest of his days dozing in a field. At any event, here it was, my newly-discovered carriage to yesterday, if we could coax it into one last foray on the open highway. The Komati and the images from prehistory beckoned.

On a fine summer morning that festive season, with only the vaguest of directions, and with the green mountains of the Mdzimba and Lugogo, Sheba's Breasts, newly washed after the storm, and clear in the already high-angled sun, we climbed into our Land Rover and set off to find the past. Rising up the tortuous, steep incline to Mbabane, my brother-in-law Christoph was driving. He understood the peculiarities of this Land Rover. The three of us were huddled on the bench seat in the front. The gear-box was screaming, the engine throbbed and the side panels set up a rhythmic percussion which made conversation difficult. But it was an African adventure. Though I didn't know it at the time, it was my first of so many that were to follow. We drove beyond Mbabane, passed the Swazi Joy Bottle Store and Take Away and on into the green, green mountain ranges of the west. At close to 5,000 feet up, all the way so far on a tarred road, we veered off right onto the dirt. I use the description 'veered' rather than 'turned' since this Land Rover seemed only to have sporadic steering; apparently it was no longer a standard feature, as in most vehicles. The engineering ensemble that once served in that role had long since had the life beaten out of it, so that the bushes, rods, steering-arms and joints had ceased to work in concert, no longer at the behest of the driver. It was easy to see, at least in the case of this example, why Land Rover described their creation as an off-road vehicle. That is exactly what it persistently wanted to do - go off-road, even though our own intention was to stay on-road.

A faded black and white sign pointed up a gravel road to the Komati and to Piggs Peak, a remote settlement in the highlands of northern Swaziland. As we hit the dirt I was introduced to a sound that was to become a familiar noise to me in Africa, the sound of an elderly Land Rover, seats, panels, doors, windows, nuts and bolts worn and slackened to breaking point with age, being driven on a badly-graded earth track which was rhythmically crossed by countless corrugations, riddled with deep pot-holes and in part washed away by the summer rains. The resulting discord was tympanic such that common speech was rendered useless above the thunderous vibrations. I soon found that, should I have wished to, I could have hurled abuse at my fellow passengers at the top of my lungs without fear of reprisal. No one would have heard me anyway. I hung on in silence. But the view through the dust and vibrating windows was staggering.

Off to our left lay the Ngwenya range, rising in undulating summits to over six thousand feet above sea level. The crest of the range was soft and lush with flushes of dark, new grass, a jade silhouette against a cerulean sky scattered with blotches of downy, flat-based clouds. In the clefts and folds of the streams flowing down its sides were lines of tree ferns - long, bent trunks with sprays of ferny foliage luxuriating from their canopy. The open slopes were dotted with short, flat-topped acacias, their shadows adding a stereoscopic effect to the mountain flanks, blots of shade in a bright and pristine emerald landscape. I was to learn later that this was the haunt of troops of baboons, Vaal rebuck and the rare blue swallow and even rarer bald ibis. Right now it looked to me like a set from Conan Doyle's *'The Lost World'*.

Beyond Ngwenya, which from its profile the Swazis had named 'The Crocodile', in the distance rose the high-reared triangular peak of Silotwane, and beyond that, blue in the distant haze, lay the ranges of Diepgezet, and Bulembu, 'The Spider', the highest mountain masses in Swaziland. It turns out that these mountains are made up of the oldest sedimentary rocks on earth, over three billion years old, a tiny window miraculously preserved from a time when the earth itself was young. It is said that among this primordial soup thrust up from the abysses of the earth's earliest oceans as hard rock are the world's earliest life forms, microscopic blue-green algae. Amazingly, much, much later, but still a hundred thousand years ago or more, these same rocks, cooked, folded, contorted and metamorphosed by the passage of aeons of time and now glassy with age, were selected by Stone Age peoples as raw materials for their stone tools, their scrapers, points and spoke-shaves that enabled them to endure in this majestic yet hostile landscape – the earth's earliest rocks with the earliest signs of life incidentally enabling our own early ancestors to survive in the world through their improved technology. Right now, our own personal cacophonic cocoon of modern technology was lurching its way further from the known world as we fought to keep it from colliding with the impressive exfoliated granite boulders that were beginning to appear along the side of the road, though at that stage understanding the exact geological context of these outcrops was less important than avoiding the imminent motoring catastrophe they represented.

Small kraals - clusters of beehive grass huts, cattle byres and mealie fields - lay supine in the summer heat, the sun reflecting off the thatch and green leaves with a silver lustre. They were seen in a moment, only to be engulfed in the dust cloud of our progress. We splashed through streams sparkling in their upland beds, surrounded by black and green wattle trees heavy with piquant yellow mimosa blossom. The wattles were an import from Australia some hundred years before, grown here to provide firewood, building materials and tanning fluids from the bark.

We whined on towards Nkhaba, our first stop. Totally incongruously, on a steep, rocky corner of the road among the dark wattle groves huddled a tiny, abandoned European cemetery, the only remains of the onetime settlement of Forbes Reef. Around 1880 a short-lived mining community had sprung up along the Ngwenya range where eager prospectors came to look for gold, only to succumb to wild Africa and to be buried in this far-flung veld. Most of the overgrown and crumbling grave-stones were illegible, but one had a marble headstone showing a winged and youthful angel staring forlornly into the far distance. It commemorated the death of a young boy who had died there, now *Safe with Jesus*, the inscription read. I wondered how many other pioneering grave-yards there were in Africa, hidden in some remote forest just like this one, forever lost and forgotten.

Emerging from the wattle into the open sweep of the upland valley again, we rattled ever northwards and further out across the unknown. At another turn, high above the Komati valley, we listed to the right, off the road and into the '*bundu*' – as Christoph said the wilderness is locally called. I remembered seeing a slim volume back at the farm entitled *'Don't die in the Bundu'* written by some colonel or another in the then Rhodesian army. It was a handbook of survival techniques and outlined how to signal to passing aircraft, how to extract water from a stone, how many minutes you might live after being bitten by venomous snakes, and which plants you could eat and which ones would kill you at the merest touch. I wondered why we hadn't brought it with us. Perhaps we should have spoken to the good colonel before setting out, but then, from what he had written, he was probably dead already. The track onto which we had turned gave way to a zigzag footpath which, without the willing cooperation of the Land Rover, we straddled. An aged Swazi man in skirt, sandals and threadbare hacking jacket leaned on his stick by the track. With feigned indifference he watched through rheumy eyes as we heaved our way past.

"Is this Nkhaba? Nkhaba?" we asked.

We tried it with a click, "N!khaba?"

He looked impassively into the middle distance.

"*Yebo*," he said after a while. "Yes".

Unfortunately '*Yebo*' can be construed in siSwati, the local vernacular, as a positive or negative response, or just "Hello." I was to learn that Swazis hate to give a negative answer and would rather concur with the enquirer than disagree, which is considered impolite.

"Is this the way to so-and so?"
"*Yebo*."
"Or is it in the opposite direction?"
"*Yebo*."

The resulting confusion is naturally overwhelming.

The footpath petered out and the way ahead became deeply gouged in a series of idiosyncratic swerving ruts drilled into the ground by the passage of decades of ox-drawn sledges laden with drums-full of water and firewood.

The back of the sledges had scored heavily and arbitrarily into the weathered granite. Like the swaying of oxen or the stagger of a drunken man, and probably the result of both, the trails weaved their way between impressive granite tors and boulder fields above the Komati. No wheeled vehicle had ever been along these bizarre, meandering paths, but we plunged onward anyway. We engaged four wheel drive. The cogs in the elderly gearbox complained bitterly at the intrusion, gritted their teeth, then ground forward with a disconcerting jangling. This, I was informed, was called '*bundu* bashing', the practice of thrashing one's Land Rover through the trackless bush regardless of vegetation or terrain. We squeezed between towering granite piles, scratching the running boards and screeching ahead. But *bundu* bashing notwithstanding, nowhere was our rock painting to be seen. In fact, we had no idea what to look for, or how big it would be. We might not reach our archaeological goal but, my goodness, bash away. This was exciting stuff. I forgot about the colonel and his scare-mongering.

I remember as a boy in Shropshire going to the Regal Cinema on wet Saturday afternoons to see the latest John Wayne western. Before the main film began we were made to watch the boring Pearl & Dean adverts, amongst which was one for Land Rovers. I must have seen it so many times, waiting for 'the Duke' to blaze away at the baddies. I even remembered the buy-line. 'In the world of go-anywhere, do-anything vehicles there is nothing to touch the four wheeled drive long or short wheel base Land Rover.' A short clip followed, showing a gentleman farmer in tweedy plus-fours and flat cap climbing into his station wagon and driving through a muddy puddle. Beside him was his well-groomed trusty sheep dog leaning out of the window. Bucolic bliss! That was what Land Rover thought their vehicles were meant to do. What we were doing, as we heaved and rolled down the vermilion and grass-green track and between the grey granite cliffs, now on solid rock, now leaning over at what felt like forty five degrees, hanging on for dear life as we crawled through dense undergrowth, was the equivalent of something extra-terrestrial. No tweeds here, just super-humans in shorts. But this was the life. I can never look at a 4x4 in a supermarket car park now, with an elegant housewife pushing shopping into the rear seat, without wanting to go over and tell her what her 4x4 is really for - hunting rock paintings in the heart of Africa. Bet she'd never tried that with her Jeep Grand Cherokee. Except we still hadn't found our paintings, and the day, and the Land Rover, were wearing on.

We pitched from one rut to the next, passed a well-kept traditional bee-hive homestead, thatched with new grass held down by a net of fibre ropes.

A Swazi lady in traditional dress stood immobile among her green mealie plants, admiring her growing crop with a beatific smile. We stopped, and after the usual courtesies, tried to enquire about Stone Age art but it seemed that in this remote and pastoral setting, incomprehension reigned supreme. We pleaded with her, making fantastical arty gestures and referring to short people who used red paint a long time ago. This only made the whole charade more obscure. She rolled her eyes skywards and pointed carefully in each direction in turn. Had she understood? Maybe there was more than one site? Maybe she was an expert in rock art? Or maybe we really were unable to express the concept of prehistoric rock art with enough conviction for her to recognise the genre. After all, in her eyes we were just another bunch of '*belungu labahlanyako*', crazy white people, in a pensioned-off Land Rover. Maybe she thought we were from the Government, or worse, from the police, which would make any willingness to understand on her part even more fugitive. We carried on none the wiser until even the sledge tracks petered out.

Grudgingly, fighting against the hostility of runnels and rocks, we turned around and laboured our way back up the slippery slope, passed Our Lady of the Mealies, still motionless among her swelling harvest, towards the crest of the hill. As we reached the skyline, we saw to our amazement what we had been searching for, standing proud above the green sward about three hundred metres away. Recoiling from the termite mounds beneath our wheels, we bounced across the veld, forcing the Rover onwards to our goal. We were not disappointed. What we saw in front of us was one of the most spectacularly-sited rock art locations I have ever seen. I have since visited many hundreds of sites in southern Africa, but this was my first one and the most memorable. Towering in front of us was a granite boulder the size of a two-storied house surrounded by lawns of alpine-like greenery sprinkled with innumerable orchids and tiny wild asters.

The boulder itself from the other direction had appeared indistinguishable from all the other tors, but from this reverse side it was very different. Weather and time had conspired to exfoliate and naturally erode the face into a huge apse-like half-dome, concave like a colossal, hollow, bowl-shaped cinema screen set in the rock. Unlike the uneven, warty, gun-metal grey of the outer surface of the boulder, this inner bowl, overhung by the rest of the tor, was a smooth creamy white. This was the twenty foot high rock canvas upon which the prehistoric artist had conceived and executed his great *oeuvre*, a painting of five full-sized eland, largest of all the African antelope, perambulating across the screen, observed in the minutest detail. To one side stood two gaunt human spectators wrapped in karosses, leather cloaks, in front of whom crawled another carrying a small bow. That was it. A single event executed in dark blood-red iron oxide paint. It was clear that there had once been more detail, for example in the area of the elands' heads and under-belly and lower limbs, probably in a stannic oxide white. This colour had long since fled, giving these magnificently rendered beasts a slightly short-legged appearance, but this in no way detracted from the stately drama of the scene. Here was a voice from long ago, clear and unequivocal in its oneness with nature, vital in primal human vigour yet shrouded in mystery. We climbed up bare-foot into the bowl, much as the artist must have done, to get a closer look. What were we to make of this ensemble? How old was it, and what human hand had so carefully envisaged and accomplished this masterpiece? It was completely tantalising, and I was completely entranced.

* * * *

We had been told of another painting site on the other side of the Komati river, and although the day was passing, we had renewed enthusiasm to learn more. Regaining the main highway, as we had by now come to view it, but actually it was the same execrably corrugated dirt road as before, we drove on. A few minutes later we skidded to a halt into a drainage furrow. In front of us opened up the most spectacular vista I had ever seen – a huge transverse gash in the mountains twenty miles across and two and a half thousand feet deep. This was the valley of the Komati. It was like an aerial view of Africa. The high mountains seemed suddenly to abandon their ambition as an upland plateau and to plunge downwards into a series of yawning gorges far below us. As they did so the landscape became by turns more wooded, scattered with innumerable beehive homesteads and mealie fields in endless terraces along the contours. These terraces had apparently been put in by the British years before to reduce soil erosion at a time when Swaziland was a British Crown Protectorate. The resulting patterns were laid out in a vast patchwork of curving parallel striations, as though the whole country, all the spurs leading down to the river, had been combed horizontally by some gigantic hand and coloured bright green where the maize had been planted. The road wound downhill like a red snake among green fields, and way in the distant haze wound back up the other side of the valley, two thousand feet or more to Piggs Peak. This was a view that summed up Africa so far for me, immense and timeless.

But timeless we were not. We had to move on. Low gears buzzing like cicadas in the heat, we crawled down the mountainside. As we descended the vegetation changed. Cabbage trees with fruits like dog's dicks stuck up in the veld, and gnarled old kiaats hung with huge medallion seeds dotted the hills. Lower down we came upon the enormous cactus-like arms of the candelabra trees, the giant euphorbias. Actually there are no cacti in Africa, but this euphorbia is the nearest thing to them.

Little children danced along the road-side at our approach, kicking their legs high and stamping their feet in the dust like miniature warriors before a battle. As we edged lower, the heat palpably increased. Finally we reached the valley bottom, where a low-level concrete causeway crossed the torrents of the Komati. The river cascaded out of a gorge upstream, the waters churning chocolate with silt from the recent rains as they surged over the boulders. A rough shelter by the bridge made from branches formed a stall where local craftsmen displayed bicoloured bowls made from the nearby kiaat copses. The heart of the kiaat tree is a rich chestnut colour surrounded by a pale outer wood, and this they carve into open bowls – salad bowls, hors d'oeuvres bowls, bowls for nuts, crisps, olives and the like – with white sides and a russet stripe in the centre. There must be endless white and brown kiaat bowls like these all over the salons of Europe. How many people, I wondered, knew the remote and wondrous place from which they had come? We stopped for a second. I offered my own advice to the carvers. "Keep your bowls open", I shouted, by way of encouragement. We drove on over the bridge and ground our way up the other side.

After labouring up the sinuous slope for what seemed an age, steam wisping from under the bonnet, we climbed the last hill into a tiny wild-west sort of settlement with hitching rails. This was Piggs Peak, named for one William Pigg who had been a gold miner here in the 19th century. We drove up the main street, a dirt strip between eucalyptus trees, grandly named 'Sir Evelyn Baring Avenue'. It transpires that the said Evelyn Baring who gave his name to this cart track had enjoyed an illustrious diplomatic career in South Africa and Kenya in the nineteen forties and fifties. Known as the last imperial proconsul, and eventually elevated to a peerage as Lord Howick, rumour has it that he was ultimately related to James I, King of England. Not only that, he was the son of one of the greatest Victorian colonial panjandrums, Lord Cromer, who seems to have spent most of his own career administering the grand financial remonstrance to recalcitrant Egyptian pashas and incompetent maharajas. Well, they both in their time might have lorded it over half the British Empire but I very much doubt they had ever bowled up their own eponymous avenue in Piggs Peak, Swaziland in a ramshackle Land Rover, fuelled with the excitement of another imminent rock painting discovery.

The flesh-pots of Piggs Peak, a one pump filling station and Dups General Store offering Joko Tea and Paderax de-worming powder, beckoned. But these blandishments failed to deflect us from our quest; we didn't linger. Lurching out of Piggs Peak we followed the tracks through an endless man-made pine forest, Peak Timbers. The tree trunks stood along the dark and gloomy forestry roads in stultifying regular European rows, mile upon mile of them. After what seemed an age, we suddenly burst out into the sunlight, into what was still then called 'Native Area', the land of chiefs, elders, runners and warriors. This was the *bundu* again. We were back in sledge track, granite and mealie country - Africa. There were many more traditional Swazi homesteads here, and the ground seemed very fertile, with endless maize fields and vegetable patches. There were many more people too, and our spirits rose to know that we had a larger pool of locals from whom we could ask directions. Most were wearing traditional dress, the men in skirts with duiker-skin sporrans and the women with glistening traditional bee-hive head-dresses. We called to a number of them with the same creative gesticulations as before, but our approaches were met with a reluctance to accept our existence that bordered on the unnerving. Despite the din, clatter and plain ridiculousness of this infirm Land Rover full of '*belungu labahlanyako*' advancing down a non-existent track between mud and thatch huts, it was as though we didn't exist. Heads turned sideways and eyes glanced away in what zoologists enigmatically describe among animals as 'avoidance behaviour', the ability of the onlooker to pretend that none of this is really happening. Only the little children, bums hanging out of shredded shorts, seemed pleased to watch us tumble by. We waved to each other and pitched on.

We kept calling out the name of the site, as given to us by our informant in Malkerns before we had left. "Insangweni," we mouthed. "Insangweni?" "*Uphi Insangweni*?" "Where is the place of the rock art? The little people long ago? *Penda libovu?* The red paint?" If they spoke this language, siSwati, at all, the inhabitants here were not letting on. It was as though they had all come down with an acute case of communal incomprehension. Only later did I learn that this cognitive perverseness had a deep and fundamental logic all of its own. '*Insangu*' is the siSwati word for dagga, Swazi gold, ganja, the green stuff, weed – aka marijuana. Worse still, '*-weni*' is the siSwati locative case, the suffix for the place where something comes from. Thus we were banging about the *bundu* far from anywhere in an ex-police Land Rover shouting "Insangweni? Insangweni?" "Where's the place with the marijuana? Where's the place with the marijuana?" You understand? "Marijuana?" "Red?"

For those innocents like me, I should further explain the extra significance of 'red' in this context. Rooibaard dagga, the Africaans for red-bearded marijuana, which has long reddish heads, is globally the finest hash money can buy and, so I am led to believe, produces the most hallucinogenic spliff you can ever toke down on. I use this jargon without any actual experience of having 'toked down' myself on such a joint, but I have it on good authority. If you don't believe me, read that well-known South African writer Herman Charles Bosman's short story called *'The 'Recognising Blues.'* In it he describes a whole pile of sub-tropical junkies totally spaced out on red-bearded marijuana.

But much more importantly, here near Piggs Peak, at Insangweni, 'the place of *insangu*', it was small wonder that these locals feigned a total lack of recognition of our progress. This was one of the key places in Swaziland where they actually grew the red-bearded magic. Swazi gold was their staple crop. They made their living from it. This is what agriculturally for them knocked all other crops into a cocked hat. The mealies were only for show. It might be totally illegal to cultivate this mind-bending greenery but that didn't prevent them from growing it on an industrial scale. They didn't need highly-trained farming extension officers from the Ministry of Agriculture in Mbabane to recommend to them what soil type to select. They didn't need fertilisers and pesticides paid for with extortionate bank loans. They didn't need 'Truck Africa' to advise them on haulage and distribution networks, nor the BBC World Service *'Farming Today'* to bring them up to date on world prices. They knew all of this already, instinctively. They had worked it out entirely on their own. They had planted it, grown it, harvested it, processed it, bundled it and circulated it by the tonne, with no help from central Government, US AID, United Nations Development Programmes, Canadian Aid, Barclays Bank Development Fund or any one of the hydra of aid agencies which targeted Swaziland on a routine basis. It was clandestinely traded in vast quantities up and down the secret places of the Komati Valley from the central Transvaal in the west to Mozambique and the Indian Ocean in the east.

And on top of all this, these people made a living from it under the all-scrutinising eyes of the ever-arresting Swaziland Police, in whose Land Rover we now sat. 'Ex-police Land Rover' cut no ice here. We were the 'heat'. And European to boot. We were no local bobby from Piggs Peak who could be bought off with a year's supply of the product. We represented a clear and present danger to their very livelihood. No wonder they were displaying signs of 'avoidance behaviour'. Rock art? A likely story! They didn't want us there at all.

Incidentally, and for the purists among you, what about the name of the site we were about to see, Insangweni - 'Where the *Insangu*' was – 'The place of Marijuana'? Historical legitimists and anthropological apologists will attempt to tell you that the name derives from a meeting two centuries ago between two tribal elders, or *tindvuna*, who met and agreed a truce in a long-running family feud, smoking a pipe of peace, a dagga pipe, to seal the treaty. Try telling that to the locals. They make their bucks from this recreational produce, and this is the very spot where they were doing it.

In retrospect, it's really a wonder we weren't done away with in some ritual murder there and then. After all, there are still enough such incidents that happen in these remote spots. You could just see the headlines in *'The Times of Swaziland'*. 'Archaeologists and farmer disappear in Piggs Peak area; police are baffled', that sort of thing. But the Swazis are by nature a very phlegmatic people. They have over the centuries retained their traditions and their independence with the conspicuously wise policy that if you give something untoward a long enough time and a wide enough berth, it will go away. This writ clearly applied here as well. So we continued unchallenged. Nevertheless, there must have been a heart-stopping moment at one homestead when, unlike their parents, the *bafana*, the little boys herding cattle, ran out and beckoned to us to follow them over a ridge, pointing alternately down the slope and then at their mouths. But they were not indicating the way to the regional marijuana storage facility to shop their parents to the authorities. They were about to show us the rock art site we were looking for, in return for which they would appreciate something to eat. Clearly, even in Insangweni, 'man does not live by marijuana alone.'

We left the Land Rover near a granite tor – no one would go near it – and scrambled after the boys. These lads were sensibly naked in the heat of the afternoon, except for a leather *lihiya*, a loin-cloth. They ran with innocent laughter down the steep slopes above the majestic panorama of the Komati valley. Far below us we could see the river like a silver ribbon meandering between imposing natural castles of granite boulders. The incline was partly wooded with broad-leaved African beech-woods, spiky flat-flowered aloes and flame acacias full with creamy catkins. A short walk away was a large rock overhang into which the children disappeared. Evidently this is what we had come all this way to see.

Clambering underneath into its shadow, as our eyes became accustomed to the shade, we saw to our amazement the walls of the grotto were covered with a kaleidoscope of paintings in many different colours and different groups.

There were whole rows of animals, of dancing men, of geometric patterns on the vertical faces and on the roof of the overhang. Large figures painted in black, arms flung out like models on a cat-walk, graced the ceiling. Tiny human figures with high shoulders and animal heads perambulated around the bosses of rock.

In the centre were depicted three groups which were by far the most vibrant. First came a painting of two perfectly observed wildebeest, an extreme rarity in the canons of southern African rock art. Beneath them danced four men. They appeared to be mincing across their granite stage, arms limply outstretched. In each hand they held a bundle of twigs. Their bodies were lithesome and elongated, as were their legs. There was no mistaking who they were. They had been drawn with very pronounced steatopygia - enlarged buttocks - which is the clear feature of the Khoi-San people, the Bushmen, who had been the indigenous people of Southern Africa perhaps for one hundred thousand years before the Bantu-speaking farmers put an end to their existence here only a few hundred years ago. To make identification certain, all four men were 'infibulated' - they each had been depicted wearing a bone pin through their semi-erect penis. This is not a modern African adornment, but it was very much the form of decoration practiced by the long-departed San. It may have been purely for decoration, or perhaps it was intended to be the mark of a tribal elder. At any event, these shadowy dancing figures must have lived here before the Swazis had arrived, and well before the long arm of cloying European morals reached out to ban the practice of infibulation, stating that 'wearing a penis pin is an abhorrent infringement on human rights'. Whose rights did they mean, I wonder?

Below the wildebeest and the dancers was the most striking group of all the Insangweni pictures – two floating semi-human figures painted in dark-umber, arms spread like diaphanous wings, their legs curling and trailing behind them. Their heads had been reduced to a circle with two elongated strands, above and below the cranium. I was to learn much later that these figures probably represented shamans, traditional San healers, in trance. In this altered state of consciousness they feel their limbs and heads become drawn out, as the *!num*, the power to heal, enters their bodies through their arms and neck, hence the so-called 'wings' shown as a series of lines leading into their extended arms. Passing into a rhythm-induced stupor, they begin to collapse physically. Mentally they see visions as they connect with the spirit world, a world full of the supernatural, a world where human and animal existence coalesces, a world where they consider they see God. As that doyen of African rock art, David Lewis Williams, was later to write:

'The great theme of Bushman art is the power of animals to sustain and transform human life by affording access to otherwise unattainable spiritual dimensions.' This is what was happening here at the Insangweni Rock Shelter. At some time long ago these paintings of shamans and animals had helped the Bushmen to see what they believed. Without knowing the full extent of the pageant at the time, we were nevertheless enthralled by its vitality and mystique. I for one was totally captivated by it, by the *bundu*, by Africa, by everything. I had added an ancient human dimension to the croaking toads and flashing lightening of my recent conversion. I had achieved my own state of altered consciousness.

* * * *

The sun was beginning to climb down the sky, and we were far from home, hungry and thirsty, but totally spellbound. We must have reeked of perspiration, dust and petrol, in our shorts and sweat-soaked tee shirts. But we had achieved our goal. We struggled back up the slope and, regaining our trusty Land Rover, gave our late lunch to the row of small boys that had now grown to a throng of eager faces. They tore into the sandwiches, ate the oranges and threw all the paper and peel onto the grass. Since they normally didn't have packaging with their food, we thought, how could we argue?

As we rolled and pitched our way back across the veld, the locals became visibly more animated. They waved us good-bye with a new-found enthusiasm, no doubt thrilled and relieved to see we were leaving without making a single arrest. One or two of the older men, red-eyed, peered at us myopically without comment. Perhaps they had been stocktaking and had already sampled the local produce with their daily dose of *insangu*.

After a long and arduous battle with the *bundu*, we reached the euphemistically titled 'main road' and banged and rattled our way down the mountainside and across the Komati drift. Thence we had the long haul up the other side past Nkhaba and Forbes Reef, with its forgotten cemetery *'Safe with Jesus'*, and on towards Mbabane. Progress was necessarily slow, and deafeningly noisy. The flaming orb of the sun dipped low and set over the Ngwenya range in a spectacular light show of crimson and rose. We switched on the lights, such as they were, and pounded onwards.

We travelled in relative silence, each with our own impressions of the day, until Christoph shouted the magic word - 'supper!' It was then that I remembered to my horror that Sue and I had that very evening been invited out to dinner in Mbabane by one of the social elite of its august 'ex-pat' community, a seconded British geologist who was also, it turned out, president of the Mbabane Operatic Society. He wanted to introduce me as a visiting academic to his circle of influential friends and co-workers in this tiny kingdom. When we had set out on our adventure earlier in the day, I had no idea how long and arduous the journey would be. I envisaged being back at the farm by mid-afternoon, having a quiet siesta, shower, change and off out socialising. In the excitement of the *bundu* it had quite slipped my mind. By now it was well past the time we should have arrived at our dinner party, and we were still hurtling down the tar road, unwashed, in the dark and the wrong side of Mbabane. I made a quick decision. As Africa was supposed to be known for its informality, that's what we would be - informal. We would arrive as we were and insinuate three people into the party instead of two. Easy.

Late by an hour for our appointment, we slewed our trusty Land Rover recklessly into the front garden of the bungalow in a smart suburb of Mbabane and, alighting in a cloud of dust, rang the doorbell. A Swazi maid answered the door, saw us, and looked away in total incomprehension. Maybe she too hailed from Insangweni. Her employer, our prospective host, pushed past her agape. At his door stood not the thrusting young archaeology lecturer and elegant wife from London, but three saddle-sore and elated people in stained tee shirts and shorts, covered from hair to sandals in dust and stinking of the *bundu*. Recognition took a short time. We made our excuses, abluted as best we could, and were ushered into the dining room. To our mutual bewilderment, around the table sat a group of splendidly attired society diners dressed exactly like evening guests at a fashionable salon in Kensington or Hampstead Garden Suburb, the men in chalk-striped suits with collar and tie, the ladies in long gowns and dancing pumps. Coming to join them was what appeared to be a slightly distasteful trio of aromatic beach bums.

A moment's silence ensued, a sort of wake for the original concept of the dinner party perhaps.

"Oh, do tell," squeaked one of the wives, to cover the embarrassment of us all. "Where HAVE you all been?"

I replied with all the vehemence and commitment I could summon.

"Africa," I said, and suddenly I meant it with every fibre of my being. Africa, my new-found world, which though I didn't know it then was to become my home and work-place for the next decade or more; Africa, with all its untamed beauty; Africa with its limitless horizons; Africa, with its ancient peoples and cultures, its shamans and spirit mediums; Africa, the cradle of mankind.

"Africa!" I repeated, warming to the juxtaposition of salon-like Mbabane. "That's where we've all been."

CHAPTER THREE

THE PLACE OF WINDS

Ex Africa semper aliquid novi.
Always something new from Africa.

Pliny the Elder

Africa: that's where I still was. That day, in the Komati Valley, Africa had beguiled me and I had eaten of the fruit of the tree, not so much the fruit of the tree of knowledge exactly, since I was still in a state of stupefying ignorance, but at least the fruit of the tree of wanting to know. It was like falling in love for the first time. I became enveloped in an invisible skein of enchantment. Just to be there, to be near the object of this new passion, to hear its sounds, inhale its aroma and understand its ancient mountains became my key ambition. I became hypnotized by the place-names which rolled around my head like an incantation – Ngwavuma, Mbuluzi, Mahlangatja, Hlatikulu. Surprisingly they were no longer alien to me so much as enigmatic and exotic, names with the power to conjure ancestral spirits, names that exuded an ancient authority and spoke of a history that held a profound secret which it became my goal to unlock.

Africa was my new love. Now I had fallen for her tantalising beauty, I was determined not to be parted from her. My tenacity was like that of an inexplicably smitten devotee. From recognising no wit of familiarity or desire the previous day, the fevered night had bred in me a passion such where, waking, everything had become part of the same obsession, evoking bizarre parallels with 'A Midsummer Night's Dream'. And in keeping with every deep-rooted devotee, it is a mood which has never left me. No matter how much of Africa I have since travelled and experienced - to the most remote horizons of the Kalahari, along the arid Skeleton Coast of Namibia, through the savannah grasslands of Mpumalanga or deep in the forests of Zululand - its landscapes and histories still captivate me as they did in the first flush of infatuation. From that moment I was resolved to break any bonds to stay close to the object of my new desire.

I suppose there was a sense in which it was inevitable that events should have brought about this transformation. I was born in a house built upon what I had as a young boy come to learn was a glacial moraine, one of many gigantic heaps of sands and gravels abruptly abandoned on the Welsh Marches by the retreating glaciers of the last Ice Age ten thousand years ago.

As a teenager I was tantalised by the fact that if I had been in the same spot, my own home, only a few thousand years earlier, which in geological time is like last Tuesday, I would have been standing on an ice sheet more than one mile thick, two thousand metres of solid ice, that had spread southwards from Scandinavia as far as North Wales. It must have been a very different world then to the one I now lived in, and yet it was so recent. In place of the cowslip-flecked meadows of today's Britain, with its village greens and cosy pubs, there would have been a polar desert, a blood-freezing wilderness of grinding ice and blasting gales. How, I wondered, had all this happened? The idea had begun an intellectual crusade for me, to understand more about the world around us and to discover our own place in it.

Through the various vicissitudes of my budding career, I had gone on asking the same question, in the various parts of the Europe and the Near East to which my calling as an archaeologist led me. Looking at the azure of the Aegean, or the dun of the Syrian Desert, I wondered if they had always looked like that, or had they too experienced such startling metamorphoses? And if they had altered to an equivalent degree, how did our emergent ancestors cope with these catastrophic and unpredictable changes of environment and opportunity?

And so in Africa, when Stone Age peoples like the painters of Nkhaba or Insangweni awoke as the sun rose over the Ice Age world of Europe, what was happening in the sub-tropics here in Swaziland ten thousand, fifteen thousand, twenty thousand years ago? What did they see in front of them? How different was their world to the one we see today. That had been my query before in the northern Hemisphere, and that that was my question now, here near the Tropic of Capricorn. Looking at the Mdzimba Mountains and Sheba's Breasts, across the intense beauty of Swaziland, what had formed this landscape? What agencies had shaped this scenery? What gave the veld this lush patina? And had it always been like that? Who had lived here in the past? How had they survived against the alarming variations and terrible odds that the world and its immense changing climates threw at them? In short, where have we come from? How did we get here? What is the meaning of life, and other mind-boggling questions? Clearly it was time for a beer!

* * * *

I have always rigidly adhered to the principle that one should never work or travel anywhere where the beer is undrinkable, or worse, unavailable, and that has stood me in good stead throughout my global peregrinations right up to the present day. My appreciation of Swaziland was no exception to this inflexible rule. As I posed these unfathomable uncertainties about our origins to my new-found family in Malkerns, to my delight out came the Castle lager. And it was not just the one either, but a whole *bierfest* of bottles which set a new tone and enlivened the conversation. Privately, I was gratified to see that my in-laws this side of the Equator had obviously developed the same basic malt-friendly policy. It was a good sign. The lager emancipated the questioning tongue, and more especially, it lubricated if not the answers themselves at least the channels through which I should look for guidance. And out of the fug of one such enthusiastic hop-driven evening emerged Johnny. I was told that I just had to see this man about these matters.

Apparently I needed a '*fundi*', an acknowledged expert on the subject of Man, Meaning and the Mdzimba. And Johnny was such a '*fundi*', or more properly, an '*umfundzisi*', which in siSwati can mean everything from a rural school teacher all the way through to, in this case, a world expert in African archaeology and environment, or at least, an expert in the current state of the subject. Johnny was my man. Johnny would know what to do next. We must phone him. In fact we must call him at once. We all rushed in a river of beer over to the telephone. The receiver was lifted and the handle cranked like starting a Land Rover with a dead battery, an experience with which I was soon to become all too painfully acquainted. The bellowing began.

"Hello, Exchange? This is Malkerns Three. No, Malkerns Number Three. Three, like Tree. Yes. No. Thank you too. Can you get me Mbabane please? One Eight Double One. No, only one Eight but two Ones. Yes, Eight. Eight as in …Ate. No, not Two, One. One One. That is, two Ones. One Eight Double One. That's it. Thank you. Lines are down? How many minutes? We'll wait. Thanks."

Communicating with the past was clearly not going to be a speedy affair. We whiled away the waiting time with more questions, amid a rising tide of Castles. Helpful suggestions flew at me. I should 'start an expedition to find early man in Swaziland.' I should 'comb the veld for traces of bygone ancestors.' Thankfully, before we could get as far as the subject of establishing a Southern Hemisphere Centre for the Study of Life, the Universe and Everything, the phone spluttered into life.

"Hello, is that Johnny?"

Through a blizzard of high-pitched static and crackling a faint voice from the other side of the world accepted the call.

I continued. "This is …"

God, who was I? Classical archaeologist? Near Eastern linguist? University lecturer? Farmer Joe's son-in-law? Drunken sot? I had a brief identity crisis.

"This is David," I said simply. "Can I come and talk to you?"

The phone clicked and whined. From far away came what I took to be an acknowledgement, though for all its clarity it might as well have been a sheep bleating down a very long pipe. In retrospect, since the lights of Mbabane were visible from the farm in Malkerns it would probably have been just as effective for me to have climbed onto the tin roof with a megaphone and bawled my request at the empty air. Anyway, the arrangement was made. Tomorrow I would meet my *'fundi'*, my *'umfundzisi'*.

* * * *

Next day we drove back up the Malagwane Hill, this time in the elegant if antiquated household Rover 90. We passed through Mbabane and continued to Dalriach, into a barrage of granite koppies, and there, nestling among the boulders and the grassy knolls was Johnny's estancia, *'Umoyeni,'* 'The Place of Winds.' We drove into a garden cultivated with innumerable proteas, an ancient family of shrubs that probably originate from some of the world's oldest flowering plants. On the side of the house next to the front door was a wall covered with prehistoric stone hand-axes cemented into a decorative panel. In Europe these axes would have been part of the Acheulian Stone Age industry, named after the village of St Acheul, a suburb of the northern French town of Amiens on the River Somme where they were first recognised as human-crafted implements in the 1790's. They were probably more than two hundred thousand years old. In Britain, certainly, such hand-axes would be revered as valuable museum pieces. In Africa, it seems, they were so numerous that they plastered the walls with them! What an introduction. I knocked on the door and in we went.

We were met by a short, wiry Scot and his two bull terriers, who, apart from looking belligerent and barking loudly, had in youth acquired the habit of serial farting - the dogs, that is. Indeed, so rhythmic were their posterial emissions that one Oxford don who later was to enjoy the hospitality of this generous ménage used to introduce them to new-comers as a kind of novelty circus act – "Ladies … and Gentlemen Pu..lease! Let's have a big hand for … Johnny and his Farting Dogs!" Was this why Johnny had called his house *Umoyeni*, I wondered?

Johnny originally came from Aberdeen but he had spent his working life in what was then the Colonial Service and based in southern Africa, until pensioned off early in the wake of Macmillan's 'Winds of Change' which in turn led to Swaziland's independence in 1968. As I just had, he had fallen in love with Swaziland and here he had settled and steeped himself in the archaeology and prehistory of southern Africa until he had become an authority on the subject. I described my background to him, amid the aroma of the coffee and the exertions of his canine helpers, and I admitted my fascination in, yet my total lack of knowledge of, the southern hemisphere and its prehistory, its history or any other of its features. I was nervous of his reaction.

Now, in the hallowed halls of British academe to which I aspired, amid the finest brains and most outstanding *penseurs* of the age, experiencing the cutting edge of research and the deliberations of the super-intelligent, I had become used to the most secretive, cavilling, and back-biting lack of co-operation imaginable. Scholars were so reluctant to let you see anything they were doing that it felt as though they were even covering their writing with their cupped hand, much as six-year olds do, to prevent you taking a quick peek at their work in case you out-shone them in front of the teacher or in this case plagiarised their life's efforts.

Johnny could not have been more diametrically different in his approach. He was the epitome of openness and enthusiasm, describing in detail his own discoveries in Swaziland, comparing notes on the rock art we had seen, and firing off a welter of references, citations and scholars we should look up. If I had been ecstatic before I had met him, I was soon heading towards a state of scholastic nirvana. It turned out that new radio carbon dating from many of the Stone Age sites in southern Africa, the ability to measure time from radio-active clocks left in the charcoal of archaeological deposits, showed that far from being 'young', of relatively recent age, which is what the academic world had until now believed, the southern African sites were old, very old, in fact much older than the equivalent sites in the northern hemisphere. This was revolutionary thinking bordering on the insurrectionist.

The sacred pre-eminence of sites in France like the Somme, and more especially in the Dordogne, whose stone tools had for a century held people's imagination as being so innovative, so advanced and so ancient was about to be challenged completely by these new discoveries. Not only had the first humans evolved in eastern and southern Africa more than two million years ago, but could later peoples also have developed their advanced technology here as well, maybe twice or three times before it ever appeared in Europe? France, and by association especially Britain, may turn out to be a mere backwater in the whole process of human physical and cultural evolution, Johnny said. Africa, instead of being peripheral, was about to become paramount in this study. But the study needed more people, more archaeologists, more excavations, more research. My hand shot up. Here I was.

In that moment Johnny and I became the firmest of friends. Completely off the top of my head I silently mapped in the empty air the outlines of an international expedition, scholars from all over the world, converging on Swaziland and its surroundings in a multi-disciplinary study of Man, the Universe and Everything. We would seek funds from all the bodies I knew. I glossed over the fact that granting committees in Europe and the USA ran on such rigid tram-lines and were so stingy that the moment anything new was suggested, it was pooh-poohed as being irrelevant, off-beat or downright ludicrous. Surely this new material we were discussing was so supreme who could stand in its way. This was heady stuff. It was also pie in the sky. I had no idea at all how such a thing could be achieved, let alone be made to work. I had totally let my enthusiasm run away with itself, the inevitable triumph of love over logic.

We paused to peer out of the picture window across the valley to the magnificent view of the mountains on the skyline. More coffee was brought. In the contemplative silence that followed even the dogs declared an interlude in their fundamental performance. Johnny motioned across to the far horizon. Did I know, he asked, that out there in those hills lay the oldest mine in the world? Lion Cavern, a small grotto battered into the side of a mountain in the far distance, was the oldest digging in creation, forty two thousand years old. Would I like to go and see it? Well, there was no time like the present. Suddenly our next African adventure was about to begin. It put me in mind of great Victorian adventurers such as Richard Burton and John Hanning Speke, the botanist Thomas Baines, David Livingstone and other intrepid worthies - wandering, reckless, fêted - who had originally investigated and described the 'Dark Continent' to an astonished Victorian world a century or more before.

* * * *

It was Dr Livingstone, of 'I presume' fame, who first suggested that the profile of Africa was shaped from sea to sea like an upturned soup plate. This may seem a simple concept now, with computerised surveys and satellite imagery, but at the time this awareness had been acquired through enormous risk and great personal hardship on the part of the good Doctor. In the 1850's this cantankerous, stubborn, over-zealous yet giant of an explorer and vicar of Christ had ventured to cross the great continent from west to east. Livingstone had faced the most appalling obstacles – mighty floods, impenetrable forests, peevish tribesmen, starving lions, obstreperous oxen, half-crazed hippos, crocodile-infested rivers and cataclysmic cataracts. Not blessed with a robust constitution, he was personally beset with every pestilence known to man, as well as some that as yet weren't - malaria, venomous ticks, chronic diarrhoea and excruciating haemorrhoids being among the most frequent. The whole expedition had been a nail-biting experience, or perhaps rather, a tail-biting experience. It transpires that every time his draught oxen gave up the will to continue and lay down in a funk, probably with good reason, he requested that his retainers bite their extremities – the oxen's, that is – to force them to move.

But he found that Africa is indeed shaped a bit like a deep upside-down dish. Starting with the sea's edge on the Atlantic west, there is a shallowly up-curving coastal region, overlooked by a sharp escarpment, itself surmounted by a mountain range. Over this mountain fastness the land falls into an extensive featureless plain – the Kalahari. This unending inland basin, perhaps a thousand miles across, is slightly concave though its vastness renders it effectively flat and it sits at a medium altitude, generally around three thousand feet, give or take a foot or two. Once across this plain, the land rises again to another mountain chain on the eastern side of the continent, then down the opposite escarpment and across the shallow down-curving coastal region on the other side to the sea, the Indian Ocean. A glance at a map of southern Africa confirms that there is a cordon of mountain ranges all the way around inside the rim of the continent – the Serra da Chela of Angola, the Khomas Hochland of Namibia, the Langeberg and Cedarberg of the Western Cape, the Klein Swartberge of the southern Cape, and then the mighty Drakensberg in Kwazulu Natal. Where we were going that morning was the northward continuation of this high upland cordon. The Swaziland mountainland, where we were bound, was part of the eastern interior circumference of Livingstone's African 'soup plate'.

Jumping into Johnny's shiny, pristine Land Rover, all smelling of new fabric and adhesive, with advantage of steering to turn the wheels, brakes that stopped it when requested and real tread on the tyres, none of which options had been selected with our previous Land Rover of Komati fame, we bit the tails of our oxen and set off. We were going to the summit of Ngwenya, which we had seen in the distance some days before.

The southern end of Ngwenya is called Bomvu Ridge, derived from the Swazi name for 'red'. And little wonder. The mountain was made up of about twenty million tons of some of the highest grade iron ore in the world, and recent mining operations had turned the earth, the trees, the roads, the grass, the rivulets and the dust clouds into a uniform and glittering deep rose madder. To give you some idea, whenever I took a group of people there later we always used to come away with shocking pink hair. The modern mine, really a quarry, was a precipitous chasm that had replaced the summit of the ridge with an equivalently inverted void. Massive trucks laboured up the steep man-made inclines out of this maroon-coloured abyss carrying the ore to the nearby rail head which had been specially built to take the rich haematite from five thousand feet down to the sea in Mozambique one hundred and thirty eight miles away, and it was thence transhipped to the steel mills of Japan. Many a Toyota Corolla, Mazda truck and Nissan Patrol has no doubt been fabricated with Swaziland iron. But this was the modern world. On a pinnacle of mountainside across the quarry, the ancient world beckoned.

Looking beyond the quarry and over the ridge we could clearly see Livingstone's bottom - the middle of the up-turned soup plate, that is. The horizon stretched away as far as the eye could see, a flat, featureless plateau of the African interior which would continue westwards more or less like that all the way to Namaqualand on the other side of the continent. It was an awesome thought. Below us was a gash in that endless tableland, the Steynsdorp Valley. Travelling from the west this would be the first evidence that the interior was about to break up. Rising above the valley towered the magnificent wooded flanks of the Ngwenya massif, deep among whose recesses lay rare and ancient trees like the rare *Podocarpus*, the yellowwood, impressive survivors of a far off time over one hundred million years ago. Pallid and sinuous among the craggy outcrops twined the roots of the rock-splitter figs, whose seeds dropping into crevices in the sheer cliffs had grown to break open the very ground in which they thrived. Rarest of all were the cycad forests, fossil survivors from the time of the dinosaurs and even earlier. They first made their appearance three hundred million years ago, when the coal measures were being laid down. They look like colossal palm-like trees, with huge cadmium-yellow cones growing vertically out of the centre of their evergreen spiky fronds. One of the three species found here on the

Swaziland escarpment was so rare it had only been discovered and named three years before our visit.

Teetering along a goat path high above the valley we reached a small depression in the hillside. This was Lion Cavern, which had been investigated a decade before as part of the pre-quarrying environmental appreciation of the area. It had been discovered that Stone Age peoples had battered away at a lode of specular iron oxide here, a highly lustrous form of micaceous haematite that was part of the twenty million tons of iron stone that made up the mountain. They had left their cobble mauls behind, and some of their other stone implements. However, this was not a recent affair. Radiocarbon dates placed this mine at close to forty two thousand years ago, before Toyotas, before Swaziland, before Africans, before history – a long, long way before history, in a period known as the African Middle Stone Age.

But these hunter-gatherers had known nothing of metals. What they were here to garner was the pigment furnished by that rich, dark, glistening ruby-coloured ore. Perhaps they smeared it on their dead loved ones, to rekindle the warmth of life. Perhaps they imagined it to be a metaphor for the Earth's own blood. Perhaps they decorated themselves with this shiny dust to beautify their bodies, or mixed it with resinous gums to paint the rocks of their sacred places. For whatever reason they had been here, and we may never know exactly what that reason was, what we can be sure of is that, in that they prized it at all, it indicates a level of sophistication beyond basic existence. They did not need this iron ore for mere endurance. It had no dietary value. It filled no belly nor satisfied any gnawing hunger. These Stone Age miners tens of millennia in the past cannot have been mere brutish survivalists. At the very least they were a people who had at some level an appreciation of the aesthetic, conceivably even of the spiritual. It was an astonishing thought, to be able to discern the awakening of art in the world of our Stone Age predecessors. I looked down at the chocolate-coloured glitter on my fingers and for an instant I touched the past, a fleeting brush with our remote ancestors. As quickly as it came, the moment was gone again. But the astonishment remained.

At any rate, I thought, coming back with a jolt to the modern world, at the very least these Stone Age prospectors were by their endeavours about to propel themselves into no less an august company than the Guinness Book of Records. I wondered what my desk-bound academic colleagues back in WC1 would make of this revolutionary discovery.

'Oldest mine in the World: Lion Cavern, Hhohho District, Swaziland.'

Perhaps it was time for some refreshment. I remembered Livingstone, my African explorer and hero, describing his own character-building thirst-quencher at a time like this.

"I have drunk water swimming with insects, thick with mud, putrid from rhinoceros' urine and buffalo dung."

Yes, quite so. There was a danger of taking the privations of exploration too far, we thought. A fine bottle of chilled Cape white wine was broached and from a hamper came some of Johnny's wife Stella's excellent meat pies. We lay back on the sweet-smelling turf and mused on how tough it could be in Africa, having to quaff one's cold Neethlingshof Cuvée Blanche neat, undiluted with rhino piss or enriching buffalo droppings.

* * * *

Geographically speaking, we were now in what was known locally as the 'highveld'. At close to six thousand feet the air was clear and exhilarating that fine summer's day, and the sky a deep ultramarine. The fumes of the wine and the rare clarity of the atmosphere conspired to induce in me a heady euphoria. The sensation is so well described in the opening lines of Karen Blixen's book 'Out of Africa.' It comes just after the bit Meryl Streep does in the film about "I haard a faaahm in Aaah-Fricaah, at the fooot of the N'.gong hyls" in her heavy, gargling singsong Dane-speak.

> 'In the day-time you felt you had got high up, near to the sun, but the early mornings and evenings were limpid and restful, and the nights were cold. The geographical position and the height of the land combined to create a landscape that had not its like in the entire world. It was Africa distilled up through six thousand feet, like the strong and refined essence of a continent.'

Wonderfully descriptive! But she was wrong about 'not its like in the entire world'. Here it was again at Ngwenya, part of a continent-fringing highland archipelago, an island arc of Afro-montane environment which probably here showed more affinities to the faraway mountain chains she was describing than to surrounding lowlands in this part of the world.

A mineral prospector's four-wheel drive track led towards the summit of Ngwenya at 6012 feet. As we bounced along the stony ridge we disturbed pairs of rare grey rhebok which leapt out of the open grassland and bounded away in a swift rocking-horse movement, curving up the long slopes with a flash of their white tails. Tiny ericas gave a hint of colour among the tufts of grass, and rufous-naped larks and cloud cisticolas flew up at our approach, piping plaintively. We stopped the Land Rover at a rock outcrop near the summit. The views were stunning. Peering down the other side of the crest we disturbed a troop of chacma baboons that had been turning over stones in their search for beetles and scorpions. They tumbled helter-skelter over the far ridge into the wooded ravine and out of sight. For me, it was truly astounding, like a view of the world that used to be.

The rock outcrops next to the track were made of a hard black quartzite. These rocks were almost as old as the very Earth, older than anything I had ever seen before. That in itself was a point of fascination. But then I noticed that all the protruding ribs of these rocks had been deliberately pounded, and large peels had been removed from them. And all around these outcrops I saw masses of chipped stone debris, chunks and broken flakes of smashed cores, and among them man-made artefacts. There were hundreds, no, thousands of them discarded in the grass all down the slope. I busily picked up pieces of tools. These were old, like the pieces we'd just seen at Lion Cavern, some perhaps even older. The key implements were long, thick-facetted blades longer and broader than my index finger, and hard and heavy. The edges showed minute use-damage. In date they hailed from the Middle Stone Age again, like Lion Cavern, but these were from the earliest phases of that period a long time ago. Johnny confirmed that the earliest date for these was possibly now thought to be one hundred thousand years in the past, or maybe even earlier, so the most recent analysis suggested.

It at once crossed my mind that in Europe blades like these were specifically associated with the first appearance of truly modern humans, but only about thirty five thousand years ago, at a time when the Neanderthal folk, the 'them' people, and their antiquated tools began to disappear and the 'us' people turned up in the caves of France and Spain, Germany and Italy. This blade technology was taken to be synonymous with that all-important break-through, the march of the 'moderns'. So, to scholars in London or Bordeaux these blade technologies are considered the key marker of the arrival of the most advanced European Stone Age communities, who in turn led the way much later to the invention of agriculture, the development of towns and cities, civilization, the sophistication of Mozart and Picasso, and eventually the modern world.

This apparent European pre-eminence has often been portrayed by some archaeologists with a descriptive iconography reminiscent of a totalitarian government hospital poster from cold-war Bulgaria. You know the type. Across the hoarding, striding towards a new and healthy tomorrow, comes a group of clean-limbed, well-nourished Caucasian super-men (and women), jaws set, eyes keenly fixed on their new future, the men tightly clutching their implements – their hammers or sickles, I mean – and the woman behind, chin up, with flaxen braids flying in the wind. In the distance, loping into oblivion over the hills are their diseased and malformed predecessors, their stooping backward glances full of the pathos of failure. And the buy line is 'Ours is the technology of the future.' Archaeologically, the concept was recently best summed up by a well-known anthropologist in his own inimitable Euro-speak about France thirty five thousand years ago, saying that it seemed as though these new and brilliant 'us' people in each valley in the Dordogne of western France were 'developing their own technological expressions with such a restless spirit of innovation'. He might have added, 'Just like we do now, isn't that right Mum.' In this Eurocentric paradigm, African cultural development was considered insignificant, very late, and not in the slightest bit important.

Yet here we were in Africa, six thousand miles away and the other side of the Equator, at a time three or four times earlier than anything comparable in Europe. This was the same technology. What was going on? It implied that even modern humans originated in Africa. This was a totally revolutionary concept that was likely to challenge all our preconceived ideas of cultural origins. And what about the scale of these events? In Europe there were a number of small caves where these precious artefacts had been found in tiny quantities. But that day at Ngwenya, I looked again across that highveld scene and grasped with a mind-boggling realization that the whole mountainside was thickly covered with this material, with these blades and the debris of their manufacture. There were hundreds of tons of Stone Age technological flotsam and jetsam right here under our feet. This was not just big; this was enormous. I wanted to be a part of this new theory of human development, to help break the mould of the European model of behavioural advance. I recognised that it might not be easy. Entrenched ideas were often difficult to alter. On my return to Britain I remember reporting this world-shattering information from Ngwenya to an audience of prehistorians in London. One professor, who had come all the way from St. Albans to hear me speak, seemed especially interested. I suggested I might show him a few of the artefacts I had brought from Ngwenya that summer's day. "Ah, there's no need to do that," he replied. "My remit only covers world prehistory as far as Austria."

He was not the only one who really could not grasp the hugeness of the idea, and the precocious nature of Africa in the Stone Age. On an occasion some years later, I was giving a public lecture in South Africa to a country club in some remote dorp or other in the Transvaal, as it was then. I waxed lyrical about the advanced character of peoples in the African Middle Stone Age and, reaching a crescendo of enthusiasm, described the remarkably progressive nature of the technology of Africa one hundred thousand years ago. At question time afterwards one recalcitrant old bat summed up the gulf of misunderstanding which lay between us when she asked, in that flat, guttural South African drawl:

"Well, Doctor," she wheedled, "'ow come if Africans 'ere were so far advanced then, they so backward today, hey?"

Oh, the pain of it all.

Back to Ngwenya. I came down from the mountain. I had seen the Promised Land and I was determined to get there. Forgetting I was a guest in a foreign country, I quite naturally turned to my new-found *'fundi'* and asked:

"So where are we going tomorrow, Johnny?"

Without a moment's hesitation, as though he had fully been expecting the question, he said, "The Ngwempisi river in the middleveld. Time for a little Acheulian."

CHAPTER FOUR

CHRISTMAS HAND-AXES

Africa is less a wilderness than a repository of primary and fundamental values;
Less a barbaric land and more an unfamiliar voice.
            Beryl Markham, "West with the Night"

So much for the highveld, whose dramatic ranges have earned Swaziland the title, at least in tour brochure jargon, 'The Switzerland of Africa.' Incidentally, we once had a visitor to Swaziland who after his stay was travelling back to the U.S. via Zurich. "That's in the Swaziland of Europe" he quipped.

Thinking of Livingstone's dinner service again, being on the eastern side of the plate Swaziland steps down from the highveld escarpment in a series of obvious yet gigantic terrace-like climatic zones towards the Indian Ocean, the plate's outer rim. There are places where you can actually peer onto the next part of the downward gradient of the upturned plate's edge. Travel down past Mbabane and into the Ezuluwini and Malkerns valleys and beyond, down from six thousand to three thousand feet or less, and this is the middleveld. The middleveld enjoys only half the rainfall and higher average temperatures. This in turn promotes a different environment, a different flora and a different human response. That was where we were going tomorrow. In fact, that was where Sue and I were heading right now, as the family farm was technically in the middleveld. Coming down the Malagwane we did a spot of low flying in the old brakeless Rover 90 and glided back to the farm. What a day!

The rural parts of the middleveld are more like the expected picture of Africa than the Afro-montane highlands. The scenery unfolds in a series of gentle undulating hills dissected by ephemeral sandy stream beds and slow-flowing rivers. Swaziland has plenty of rain which falls on the mountainous western ranges during the summer months. These feed the eastward flowing streams like the Ngwempisi, Mkhondvo and Ushutswana which in turn swell the major seaward- trending rivers of the Komati, Mbuluzi, Usutu and Ngwavuma. This must have been a land of plenty in the ancient past too, for even today in a reasonable year the grazing is good and it must once have teemed with game. Now alas only a few wild animals like the duiker, a shy but resilient antelope, survived in the middleveld. It was here that the Swazi nation settled almost two hundred years ago, where the grass was nourishing enough for their all-important herds of cattle, unlike the sour unpalatable grasses of the western highlands where acid soils were leached by too much rain, or the low-lying east riddled with tsetse fly and ticks which brought disease and death to their livestock.

As the mountains give way to the middleveld the scenery broadens into natural open grassland with thickets of thorn bushes. Along the sandy courses of the smaller streams grow wild date palms and the fever trees, with their surprising acid-drop lime-green coloured bark and delicate leafy tracery. Clusters of beehive-hutted homesteads punctuate the landscape with their tree-trunk livestock byres. Stalked by white egrets, herds of African piebald cattle, sporting huge horns, wend indolently across the corrugated dirt roads or stand in groups in the shallow brooks, shaded by overhanging Natal mahogany trees and stately sycamore figs. It promotes a soporific lifestyle, where in today's world time runs at half speed. Here in the middleveld we were to find many indications of the ancestral time in the ancient past, innumerable sites that charted the passage of human groups way back to the beginning of prehistory, longer than Europe had ever know by a long way. That was to be our next adventure.

※ ※ ※ ※

There are exceptions to this rural middleveld idyll where time stands still. The Malkerns valley, centre of middleveld commercial farming, was such a one. We arrived back at the farm to find that evening the family had arranged a party in honour of my wife and I, and preparations were in full swing. I was to be presented to the neighbours. In expectation of that, they had killed the fatted calf, or at least, the fatted lamb, a whole lamb roasted on a spit, naturally with the full assortment of alcoholic accompaniments.

The local farming community arrived couple by couple and I was duly pushed forward to meet my in-law's friends. They were all intrigued to meet the man who had finally carried off Farmer Joe's daughter, Sue, rightly considered to be a well-known beauty. Most had that robust attitude of 'Well done, son. Someone has captured her heart at last.' Variants of that, slightly more obliquely praiseworthy perhaps, included things like, 'Thank God! We thought she was never going to get married.' When I announced that we had just discovered that Sue was pregnant with our first child their warmest congratulations knew no bounds, though I thought I noticed one or two of the more conservative matrons slyly counting their fingers. One such harridan, who apparently knew everyone's business, seemed to take a genuine interest in my career, so I made the mistake of mentioning the blades of Ngwenya.

"So, you're an archaeologist. How novel. Still, you've got your lovely wife to support you haven't you."

I soft-pedalled the Stone Age after that, in case they thought I was even more of an impecunious eccentric than they had already imagined. It reminded me of a conversation at a dinner party in Chelsea some years before. When I announced that I was an archaeologist to the lady sitting next to me, she cried, "Oh, how absolutely fascinating. And what do you do in real life?"

The Malkerns party progressed. I was introduced to the local *femme fatale*, who it seems had been married four times already and was on the prowl again. Maybe she was hoping to reach double figures but by the time I met her, her good looks were already fading. She was Swaziland's equivalent of those vapid sirens that turned up in Kenya during the twenties and thirties, like Diana Caldwell or Beryl Markham. I silently transposed the relevant adage in my mind – 'Are you married or do you live in Swaziland?' She rested a bony hand suggestively on my arm and whispered some slurred advice on matrimony. Mercifully I couldn't hear what she said above the hubbub, which was perhaps just as well in the circumstances.

But the lamb was cooking well. My brother-in-law Christoph was an expert in the art. He pointed out that the heat had to be placed under the legs, not the rib cage. That's where the densest muscles were. Suddenly, he lifted his empty glass and balanced his on his head. At first I mistook this for the beginning of some acrobatic ritual that bar-b-cueists performed in this part of the world, something like crossing the line, maybe. Fortunately it turned out it was only a signal for yet another krug to be broached. Nothing loath, Christoph and I made a further frontal assault on the next Castle, like the honest yeomen we were.

Finally, the roast meat was ready. With a tool kit reminiscent of a service bay at a small garage, the carcass was attacked and wrenched from the spit onto a table. Christoph tore a piece from the back fillet and handed it to me. It was delicious. He then set about carving the legs for the assembled guests who were already lining up with salad-filled plates. It was now that I learnt an important lesson in African lamb bar-b-cueing - if you cook, don't cut. Christoph had been nursing and nurturing this beast from raw to perfectly done for four long hours, from sun-down to night-time, in the heat of the fire. Thirsty work, no doubt, which had required constant libations of beer; his glass had been balanced on his head many times in the course of the enterprise. The conversations and discussions around the coals had kept him amused the whole time, but now, coming away from the fire into the comparative cool of the veranda, the accumulation of Castles began to exact their toll. His carving skills took a downward lurch, as he sliced into the legs, the shoulders and his own fingers by turns, oblivious of the difference. Other workers took over, and the party turned out to be a great success.

I was introduced to another wonderful tradition that evening. Towards the end of the party Christoph motioned to me that we should relieve ourselves, which we did in the penumbra at the edge of the lawn. The sky was a bright canopy of stars in the jasmine-scented night air. What a marvellous experience, that comfort stop under the heavens. It was a time for expansive gestures and broad philosophic declarations. In mid-stream Christoph leaned unsteadily towards me, a glass in one hand and his personal equipment in the other.

"Davie," he admitted, "I was prepared to get on with you for the three weeks you were here whatever you were like, you being a hairy archaeologist and that. After all, you did marry my sister. In any case, it was only going to be for three weeks. But you know what? You're a really good bloke. That's really great. Cheers!"

I felt I had that night passed another important test - my induction into the All-Swaziland serial beer-drinking fraternity.

* * * *

Early the following morning, Johnny drew up in front of the farm next to the wreckage of the night before. Taking stock of my personal situation I realized I had been assaulted by the hop. My head pulsated and my mouth felt like the bottom of a parrot's cage. But the day had dawned sunny and warm, and we were going back into the distant past again. Johnny's shiny time machine awaited. After a couple of alka-seltzer, a cold shower and a jug-full of water, I was ready to go, off to the Ngwempisi, wherever that was. We swung out of the farm, through the little settlement of Malkerns, famous for its pineapple and citrus cannery, past the mud and corrugated iron shanties of the cannery workers, past the local bottle store, nexus of the previous night's mischief, and out onto the dirt roads of the middleveld. We clattered over the low-level bridge across the Great Usutu River, still a slender brook at this point, and then turning under the lee of Ntondosi Mountain we plunged into the heart of rural Swaziland. Huge erosional gullies slashed the deeply weathered bedrock in a series of livid scars, and between them lay the thatched homesteads of Swazi farmers surrounded by vivid green mealie fields.

Further on the open grassland gave way to tangled scrubby woodland, a jumble of seven or eight species of thorny acacias and various bush-willow combretums with their brown winged seeds like dark flecks against the lush green foliage. We crossed a number of boulder-strewn dolerite dykes, huge walls of volcanic lava from beneath the Earth's crust that had been squirted into faults in the granite hundreds of millions of years ago. These boulder fields were like piles of gigantic chocolate-coloured horse manure running in chains across the crown of the jade countryside. Next to them the weathered granite in the road cuttings was a pink mass topped with waving grasses. The granite had been so heavily de-natured over millions of years by rainwater dissolving some of its minerals and air oxidising others that it literally crumbled to dust at the touch. From one of the dolerite ridges we looked down into the majestic valleys of the Ngwempisi, the middle Usutu and the Mkhondvo rivers laid out beneath the Mahlangatja hills. It was a spectacular sight. Somewhere in that huge coliseum of wooded scenery lay the confluence of the three rivers, which was where we were heading.

On a low-angled slope above the course of the Ngwempisi we stopped the Land Rover by the side of the dusty track and walked into a fallow mealie field. The clods of soil had been roughly turned over by an ox-drawn plough the season before, and dry mealie stalks lay scattered among the furrows. Johnny started to walk, eyes down, across the plough land. He looked intently at the ground. We followed suit, quartering the area in a survey of the field. Some years later, standing with some Swazi rangers while my own survey team was doing a similar thing in another part of the country, searching for stone tools on their hands and knees across the veld, I heard Mnisi, one of the rangers, ask the other "What have these *belungu labahlanyako,* these crazy white men, lost this time?"

The plough-land over which we were walking was quite some distance from the Ngwempisi river, left high and dry some hundred feet above the river's present course when it had cut down in its bed in some previous climatic era. And on this surface were hand-axes from the Early Stone Age.

Johnny picked one up and passed it to me.

"There", he said, "what do you think of that?"

Holding this stone axe, the size of a paperback book, in my hand was an electrifying experience. They had been named 'Acheulian' hand-axes in Europe because of the French connection, but the ones in Africa were very much older than those found in the gravels of the Somme in France. It was impossible in these open plains to know exactly when this axe had been made but the earliest hand-axes in Africa, from Kalambo Falls in Zambia, had been dated to almost 1.5 million years ago. It was quite possible that this axe too was well over five hundred thousand years old, a stone implement that had been carefully fashioned by one of our human forebears to butcher the prey they had hunted before dividing the meat between their family members. The original detail of the incident had long since vanished and all that was left now was this discarded cutting tool. But looking closely at the carefully chipped facets of the double-sided axe I now held in my hand immediately brought to life the bright noonday of that time long ago.

Even as I was pondering this, Johnny had already found several more. They were everywhere, once left abandoned on this old river terrace and now unwittingly exhumed by the farmer's plough-share. The profiles were oval, lance-like, rectangular, all shapes, but together they shared that careful workmanship, a flaked sharp edge around their circumference which had enabled these people to survive and thrive in what must have been a frightening world, a world full of terrifying predators and scavengers with whom our forebears were in direct competition – lion, leopard, hyena. Back in London there were those that spoke knowledgably about hand-axes and their morphometric ratios, percussion angles, flake efficiencies and cumulative frequency graphs. In Africa, where it all began, this was no brick-built laboratory. This was the genuine article. Once again I was captivated by the reality of place, which gave an infinitely more vibrant context to it all. I was desperate to know more.

Not everyone appreciates the uniqueness of it, or hears that primal call, or feels that nearness of the past in the same way. I remember some years later being in the same area. I was conducting another survey with some British and American volunteers near the confluence of the Ngwempisi and the Great Usutu. Along the river grew pristine African savannah. It was winter then; the gnarled buffalo thorns rose above the long, dry grass and the elegant fever trees were in bloom, filling the air with the sweet yet sharp perfume of mimosa blossom. Vervet monkeys ran through the branches of the boerbean trees, heavy with crimson nectar-dripping flowers, and in the distance a small group of sand-coloured impala grazed in a clearing. As we walked, from a spike-thorn bush right in front of us three male waterbuck with elegant curving horns suddenly sprang forward and crashed into the *phragmites* reeds at the river's edge, and as we approaching the rushing water a gigantic crocodile slid from the granite boulder where it had been warming itself in the morning sunshine and plunged into a nearby pool. The scene was the epitome of wild Africa. Scott, one of our U.S. volunteers, an accountant with Mercantile Credit Inc., Gaithersburg, Maryland evaluated the scene.

"Why, this is just like the Potomac," he opined.

Well, if there's one thing it was not like it was the Potomac, nor anything else in the northern hemisphere I had ever seen, come to think of it. This was indeed Africa, where it had all begun for mankind.

Johnny kept finding more and more hand-axes. This was like Christmas. But just a minute, this was Christmas. I suddenly remembered the initial reason for our journey to Swaziland. Our journey into the past would have to be halted briefly while the proper offices of the present Yuletide were observed.

* * * *

Back on the farm in Malkerns the end of the year had arrived. It was time for the staff and workers' Christmas party. An elderly ox had been shot and slaughtered, not without some difficulty it seemed. But it was now in kit form, hanging from hooks set up in the shade of a flamboyant tree, the maroon and white marbled flesh against the orange flowers and light green filigree of the leaves creating an oversized and bizarre polychromatic mobile. A couple of Swazi amateur butchers were hacking at the meat with *pangas*, watched by farm workers in blue overalls standing about looking sagely and expectantly at the impending victuals. Some wore hard construction hats and gumboots, even in this heat. I was told that it was like a badge of office. They had once worked down the deep levels of the gold mines on the Reef in South Africa and curiously, despite having to live with apartheid there, they were still proud of the fact.

A huge bonfire had been built and covered with what looked like the metal equivalent of a five-barred gate, and the ox roast began, piece by piece. With their magnificently strong and pearly teeth the Swazis preferred to eat the tougher cuts of meat, which they reckoned had a far more beefy flavour. In the local Malkerns Valley Butchery up the road you could pick up a whole fillet steak for next to nothing. Nobody wanted fillets – not tough enough, and no taste. What a place to be, I thought.

Then the *tjwala* appeared, the local beer, brewed by the women in terracotta *timbita*, huge spherical terracotta beer pots. The beer used to be made from sorghum, and indeed the best local beer still was, I was assured. But this homemade beer was made from fermented maize flour and looked a frothy, murky white in colour. Each man, seated on the ground, took a deep draught and passed the pot to his neighbour with a satisfied sigh. More meat and more beer appeared and the management handed out small gifts. Each man took his package with a peculiarly Swazi gesture, knees bent, bowing slightly and putting his left hand lightly against his right wrist as he extended his hand. I thought it both charming and slightly obsequious by turns, but I was informed and later observed that it was the custom throughout Swaziland, not so much to show humility before your benefactor, but to demonstrate that you have no weapon hidden in your left hand with which to stab him to death. It reminded me that the AmaSwazi were a warrior nation, like their close cousins the AmaZulu, and their history was written in blood.

As if to emphasize the fact, the assembled workers rose at that moment to dance. In a clear voice a chanter set the tempo, and with stamping feet the others followed, each with both hands outstretched holding a knob stick vertically at arm's length in place of an assegai. The measured rhythm strengthened and the feet pounded harder, gumboots notwithstanding, raising the dust and making the earth tremble. They advanced and stamped, now once, now again, the beer seemingly emboldening them to battle. The women, choral with *tjwala*, were clapping wildly, ululating and intoning what I took to be a traditional war chant, like in the film 'Zulu', urging their men folk on to victory. However, it turns out that this was not the Swazi equivalent of 'Men of Harlech' but rather an extemporaneous stream of abuse aimed at impugning each male dancer's sexual abilities.

The women sang with a dissonant yet surprisingly appealing harmony, which certainly had the desired effect on the would-be warriors. The more agile among them adopted wide-eyed, bellicose postures and performed cartwheels in front of the other ranks, who stamped the more furiously at each turn. Being myself of Welsh descent, I thought that if I had been with the 24th Regiment of Foot before the Battle of Rorke's Drift in 1879, standing at that mission station alongside Michael Caine and Stanley Baker, and listening to the AmaZulu and their war cries, just like this, my blood would at once have turned thin and chill in the face of such a level of terpsichorean commitment. It put me in mind of Wellington's comment about his own troops during the Peninsular War:

"I don't know what effect these men will have upon the enemy, but by God, they frighten me."

That's the effect these AmaSwazi had on me. I was glad I was on the right side on this occasion.

* * * *

The staff Christmas junket over, the family trooped into the chintz drawing room for a couple of gin and tonics before going out to the carol service at the Malkerns club. Malkerns Country Club, members only, was reminiscent of one of so many such temples to Britishness scattered over the one-time empire and built to isolate would-be colonialists, at least temporarily, from the indigenes with whom they worked. Outside, in darkest wherever it was, were wild Pathans, or Fuzzy-wuzzies, or Matabele, but inside, the bar would be decked out in an attempted collage of a 19th century Kentish coaching inn, replete with pewter tankards, oak beams, horse brasses and willow pattern plates. The Malkerns version wasn't as bad as that, really. It was rather a mixture of farmers' pub, village hall, tennis court and bowling green. In place of Double Diamond, Bass Ales and Gordon's adverts there were the more local Castle, Lion Lager and Old Buck Gin. And it had its sprinkling of Afrikaner cattle men and Swazi athletes too, so it escaped the real faux-Westerham or Tonbridge kitsch.

The club attracted that inescapable cast of characters that somehow fetched up on such far-away shores - disgraced army officers, failed accountants and short-contract car mechanics posing as engineering consultants. And Swaziland had its share of remittance men too - dipsomaniac lesser sons of well-heeled county families who were kept out of the way in some obscure ex-colony with a monthly cheque and told not to come back, or the funds would dry up. The Malkerns edition was the son of a well-known brand of British sweet makers. He and his wife were usually to be found rather 'tired and emotional' as early as ten o'clock in the morning.

Anyway, here we were in The Club. My brother-in-law introduced me to the bar staff. So that I could buy my own rounds I was enrolled as temporary 'country member'.

"What kind of remember is he?' cracked one of the bar men, with a leaden Liverpudlian leer. I wondered how often he had said that. He was here as the temporary motor-pool foreman at the canning factory next door while my new family had been here forever. They had even built the club in the first place, and on my father-in-law's land too. I felt an honorary Swazilander already.

While the family made its rounds of the same people we had seen at the lamb roast, I fell into conversation with an impecunious British ex-cavalry officer sporting a monocle who sat disconsolately at the end of the bar. With an accent more fitting for the Officers' Mess of the Poonah Light Horse than rough Africa, he was willing to pour out his woes to anyone who would listen. His brother had made a fortune making children's push-chairs, he told me in all seriousness. He assured me that he himself had been about to hit the jackpot for a large, undisclosed sum by advising a certain Middle Eastern potentate on his stabling requirements. But he had had the rug cruelly pulled from under him just before he got there, he said. At the critical moment, the shah had been cruelly deposed and his successor, the Ayatollah Khomeini, seemed not to be in need of the officer's equestrian expertise, so here he was, marooned and destitute in this tiny African kingdom. His baby-carriage sibling had disowned him. I bought him another double scotch, just for maintaining the stereotype.

The carol service began. A pensioned-off piano honky-tonked the opening bars of a well-known Christmas melody and with halting voices the beery congregation joined in. It had nothing of the flourish of the workers party and dance of the afternoon but the ex-patriot merry gentlemen and ladies here gathered did justice to the 'Herald angels' and the 'Oh come, all ye faithful.' One or two other carols were not so hearty, but then I suppose it would be quite difficult to give a convincing rendition of 'In the bleak midwinter' when the temperature outside was plus thirty degrees centigrade.

I looked through the window into the warm blackness of that silent night and thought of all the many prehistoric peoples that had ranged over this land, in the time before the club or the cannery, in the time before time began to be counted. I had learnt such volumes in such a short space and I couldn't wait for the next episode, the lowveld and the true African bush. So much more to discover – 'Oh tidings of comfort and joy,' I thought.

CHAPTER FIVE

LION OF THE NATION

Groups of Bantu-speaking people migrated into south-eastern Africa in the late fifteenth century. Leading one of these groups was Dlamini, founder of the royal clan of the Swazi.
  Hilda Kuper: An African Aristocracy

It was an extraordinary experience for me to welcome in Christmas Day wearing a pair of shorts. Outside, the clear, heat-perfumed air was filled with the incessant call of Cape turtle doves - 'Work harder! Work harder! Work harder!' - so loved by filmmakers as the sound of an African morning. The sun was already bright and hot, and the garden was awash with light reflecting from the cascades of purple and magenta bougainvillea and orange cannas, and spilling off the giant elephant ears. When I was young, in the Marcher country of Britain, the weather at this time of the year resembled nothing so much as the song from the end of *Love's Labours Lost*, 'When icicles hang by the wall,' especially the bit, 'When blood is nipp'd and ways be foul.' However traditional those frost-bound, sub-arctic Yuletides long ago had been, I finally had to agree that good cheer at minus fifteen degrees paled into insignificance when compared to this year's cocktails and smoked oyster canapés around the pool and balmy evening parties outdoors. I decided there and then that I didn't need my blood being nipped anymore. White Christmas or no, 'yo-ho-ho' in the torrid sub-tropical heat would suit me just fine. Being only a hundred miles south of the Tropic of Capricorn, the sun in Malkerns shone more or less straight down the chimney on Christmas Day and lit up the empty grate. In the Salopian tundra, we were lucky to see the sun shine at all, summer or winter, let alone down the chimney.

I continued my morning reverie with a walk beyond the house and into the cultivated lands where I was immediately surrounded by pineapples. Stretching out as far as the eye could see, blanketing one side of the broad, shallow valley and up the farther slope were serried ranks of yellow-grey bursts of spiky leaves, in the centre of which, at about knee height, sat a monster green pineapple. These were cayenne pineapples, specifically grown for canning. The metallic-silver sheds of the canning factory were silent and still this Christmas morning in the dewy post-dawn haze. Normally the factory hummed with activity, belching out black smoke from the boilers that heated the juice before it was sealed with the fruit in tins. Malkerns was the only canning plant left that still selected its sliced pineapple by hand – 'fancy grade' tinned pineapple rings to garnish the gammon steaks of boarding houses from Bridlington to Bournemouth. I contemplated the difference between seeing them growing here in the shimmering heat as opposed to being at the receiving end of the food chain in the shivering cold.

Looking beyond the valley into the folds of the distant hills, where once there would have been granite koppies and green upland grazing there was now a covering of man-made forests of pine trees so dense that they blended into one sombre backdrop of misty Prussian blue. In its day this had been the largest man-made forest in the world, providing fibre for the pulp mill at Bhunya away up the Usutu valley. Sometimes when the air was really still, the locals said, you could smell the fetid cabbage aroma from the mill even here at Malkerns.

Off to the right towards the highveld the forest climbed a steep plug of rock and where it had recently been half-felled along the skyline the silhouette of the remaining trees gave a passing resemblance to a Japanese silk painting. Wheeling around I could see the unmistakable emerald green of the twin peaks of Sheba's Breasts, pert and graceful in the soft morning light, and beyond that the outline of the high ranges of the eastern escarpment, Swaziland's western boundary. There was an unmistakable magnificence to the scene, especially knowing what I now knew of this crucible of human existence.

Christmas Day and Boxing Day passed in a fog of parties, introductions, toasts to absent friends, and a spot of bass fishing in the dam on the farm. And of course we had to keep refortifying our Castles. Being a good trencherman I had no difficulty keeping up with the steady stream of krugs aimed in my direction. The day after that, however, was also declared another public holiday. While we waited for the next adventure into the past, there was time for a little local contemporary sight-seeing, and someone suggested that we should go and find out what was happening at Lobamba, beneath the shadow of Sheba's Breasts. This was where the paramount chief of Swaziland, Sobhuza II, had his traditional homestead, and today was the day of the *Incwala*, a time-honoured annual festival of dance and general revelry, I was told.

Up to that moment I had not really thought too deeply about the Swazi people as a nation or a culture, but rather had tacitly accepted them more as part of the scenery, the ubiquitous mealie farmers along the roadsides. That was all about to change.

My other new-found brother-in-law Paulo, down from his work in 'The Republic' as South Africa was called, offered to take us to the *Incwala*, and with this in mind he had cranked up yet another farm Land Rover from the implement sheds. The blue one had phoned in sick with breathlessness and lower back pain after the rigours of the Komati. This 'new' one turned out to be a grey heap of certifiable decrepitude. It had no cab, no roof, no doors, no tailgate, just a chassis with an engine, four wheels (I checked), an optional windscreen, three torn seats and a pay-load space. This was a 'bakkie', the local name for a pick-up truck. When Paulo started the engine, what was left of the vehicle's panelling began to gyrate alarmingly, even at rest. He kicked the tyres, more to ensure they were all there than to test the pressure. Tyre pressures were a servicing luxury long ago dispensed with. We clambered in and the remnant component parts began to jolt off violently in a semi-random forward movement. I grasped the side of the windscreen, watching the open road pass under my feet through a missing panel in the floor. I had really enjoyed this Christmas, and was just speculating if I would ever live to see another when my father-in-law shouted after us, "Don't worry, it's only got to go five miles up the road to Lobamba." I seriously doubted it would make twenty per cent of that, let alone bring us all back again. I noted that at each right hand corner Paulo automatically leaned over and grabbed the passengers to his left to stop them being centrifuged into the road.

But soon we were on the open highway, with the wind in our hair and a song in our hearts and our hearts in our mouths. Apart from a penchant for using the whole road, crabbing its way across both sides of it, this apology for a conveyance was going as well as can be expected, and thankfully, though running flat out, that was not very fast. Two of the gears had gone AWOL a long time ago, which regulated the tempo somewhat, though it did lead to a severe disadvantage on any upward slopes we encountered. I prayed that there would be no downward slopes. After what felt like an hour we arrived at Lobamba where this Solihull gorgon shuddered to a halt, coughing convulsively with pre-ignition. I self-consciously ran my tongue over my teeth to check that all the filling amalgam was still there. Then I looked up and saw for the first time the most startling sight – it was Rorke's Drift all over again, but this was no Christmas party on the farm. This was for real.

Standing in front of us, not three hundred feet away, were thousands of Swazi warriors in full traditional dress, row upon row of them, spread out *en echelon* across the fields. They were armed to the teeth with broad-bladed assegais, evil-looking triangular-headed battle-axes and a variety of knobkerries - wooden cudgels like Irish shillelaghs. Naked from the waist up, they each protected themselves with a long, leather war shield fashioned from the hides of their piebald cattle, in brown and white, or flecked black on grey, or pure white, or black with white patches and so on. These were the insignias of the Swazi *impis*, the king's regiments, demarked by age and shield colour. They were murmuring like a swarm of angry bees, just like Cetshwayo's *impis* in front of the biscuit tin and mealie bag defences of Rorke's Drift. Remember that famous line in *'Zulu'* when someone runs down the slopes of the Oskarberg into the fortified Swedish mission screaming, "Zulus to the South West! 'Fasands of 'em!" as the Zulu regiments came peeling round the corner? Well, that's what it felt like now, and here was I, 796 Private Price Williams, right in their path, unarmed apart from a clapped-out Land Rover. Where was Michael Caine when we really needed him, I thought, feverishly. Maybe I could put his off-screen Cockney gag to the test: "Don't you throw … those bloody spears …at me!" I could hear the sepulchral voice of Colour Sergeant Frank Bourne (aka Nigel Green) intoning in my ear, "Look to your front lad. Mark your target when it comes." "But what with, Colour Sergeant," I mouthed?

Just a minute! These *impis* weren't looking at us at all. They were looking intently in the opposite direction towards a vast cattle kraal, the *sibaya*, the King's cattle byre, whence they had been summoned for the *Incwala*, the most sacred moment in the Swazi calendar. The *Incwala*, I later learned, was a deeply significant, multi-layered pageant heavily charged with historic protocol and power. It was the reaffirmation of the nation's loyalty to its ruler, expressed in dramatic rituals associated with Swazi kingship. It was a panoply of economic and military rites of passage allied to the mid-summer solstice and the phases of the moon. It was the crucial first fruits sacrament of this pastorally dependent people. And it was the supernatural link between the King and the mystical powers of nature, in particular to deliver rain for the growing crops. Today was actually the last day of a ceremony which had been going on for almost a week already, with a preamble a week before that, but it was the only day to which outsiders were invited, to pay their homage to the re-born King and to express thanks for the munificence of his reign.

As we approached the outer entrance to the Royal *sibaya* we were able to press our faces between the upright tree trunks of the perimeter of the kraal. Inside in the far distance stood the King himself, His Royal Majesty Ngwenyama Sobhuza KaBhunu, King of Swaziland. He was seventy six years old and at that time was the world's longest ruling monarch. He was wearing the special *Incwala* regalia of a pectoral of black and white cow's tails on his chest and a dramatic headdress of iridescent feathers. Here beneath Sheba's Breasts he stood, the living metaphor for the Nation, successor to an endless lineage of kings going back into the mists of time and head of the emaDlamini clan, the clan from which all the kings of this proud people had emerged.

The early history of the kings of Swaziland is concealed in the intricate and distant past, somewhere to the north of the present city of Maputo, in Mozambique. The genealogies of the wandering emaDlamini dynasty that have been offered by Swaziland's worthy historians make about as riveting a read as the 'begats' in the Book of Genesis. But without question, the status of modern Swaziland as an independent kingdom owed everything to the skill, patience, diplomacy, cunning and shrewd judgement of this stooping old man who we could now see between the sticks of the royal byre - Sobhuza II.

Born in 1899 to Lomawa Nxumalo, a wife of Ngwane V, he was originally named Nkhotfotjeni, the lizard, on account of the fact that his father Bhunu was skulking about among the boulders of the Mdzimba mountains dodging the Boers who wanted him on some trumped-up charge of murder. Bhunu actually died during the *Incwala* of Sobhuza's birth, the 'first fruits' ceremony celebrated in December 1899. He was in his early twenties. It is generally acknowledged that he was poisoned by his witch doctors. And Bhunu's own father, Mbandzeni, (Dlamini IV), is also thought to have been poisoned during a 'first fruits' ceremony ten years earlier. He had reached his early thirties. Ironically, it turns out that for both these kings the *Incwala* was more like a 'last fruits' ceremony. Swaziland must have seemed a fairly lethal place for a young monarch in those days.

Knowing what I know now, it must have come as something of a relief even for the present king, Sobhuza, to know that he was still alive on this last day of *Incwala*. The event had been, actually and metaphorically, a testing time for him, since part of the covert rituals earlier in the pageant would have involved mysteries of personal denunciation and necromancy.

The exact moment of consummation of *Incwala* is bound up with complex and ancient astrological auguries. Preparations begin two weeks prior to the climax of the ceremony, on the night following the day of the dead moon, when the sky is black and the darkness pregnant with foreboding. As night follows night thereafter, and as finally the sun sets on the summer solstice, the most sacred traditions are enacted in total secrecy, the warriors chosen to keep these observances chanting songs about the king's marriage to his ritual wife, about the return of the spirits of the ancestral cattle from the royal graves of their predecessors, and about the ghosts of the old royal house of the emaDlamini.

With the first fingers of dawn slanting over the mountains on the penultimate day, the king must walk naked except for an ivory penis cover before his own regiments and before his royal wives - weak, helpless, reviled – the moment when his people ritually condemned him to oblivion. His reign is over; his supremacy gone. But then, from inside his consecrated hut, in the persona of the 'new king', he responds by spitting powerful medicine to the east, at the orb of the rising sun, and then at the light growing in the west, breaking off the old year and ushering in the new. By such a supernatural rite he regains his strength and his virility, and with it his kingly power. He eats of the first fruits of the fertile earth, and like those crops, he is himself reborn.

The clandestine assembly sings the anthem of *Incwala*, *Ingcaba kangcovula*, to welcoming back their monarch into their midst:

"Here is the Inexplicable, Our Bull, Father of Our Nation, Our Lion, the Great Mountain!"

There is a strong similarity in the *Incwala* ceremony to the annual dying and rising kingship rituals of the ancient Near East, in which kings literally were perceived to expire and then were restored to life, as the year is renewed, as the earth itself is regenerated. The king is believed to personify these transmutations. In some cases they may even be killed and replaced. Among the emaDlamini, with powerful '*imitsi*', the herbal concoctions being administered by the royal witch doctors during these sacraments, there could be no better time literally to do away with the old and bring in new blood. In fact kings of five generations before Sobhuza died like that, in their prime of life. Disquietingly, even Sobhuza himself was often heard to maintain that no one ever died a natural death. It is always a death by sorcery, by witchcraft, by doctoring.

But we who were gathered at Lobamba that day could testify that King Sobhuza II had survived this current *Incwala* and would live to lead his nation for yet another year. His living presence testified that the rituals of *Incwala* had been appropriately conducted. Incongruously attired foreign dignitaries gathered to stand in the cow dung nearby to witness this rite of passage. The American Chargé d'Affaires stood with his *aide de camp* in a natty lounge suit purchased from 'Cohen's Outfitters' on Capitol Hill. The British High Commissioner turned up in what appeared to be a Victorian naval officer's uniform with brass buttons, epaulettes and a cocked hat. He'd obviously left his frigate anchored up the hill in Mbabane. A variety of other diplomats were in attendance, in particular a contingent from Taiwan, who were seeking votes in the United Nations.

Meanwhile, the aged chiefs of the nation stood loyally by, wearing black head rings on their bald pates, while the impis paid homage, all dressed in their brown-patterned cotton skirts and duiker skins. The praise singers exalted their leader, while the regiments shouted in unison, *'Bayethe! Bayethe Ngwenyama!'* – 'So let it be, Great Lion!' Sobhuza acknowledged their good wishes with a regal wave. The nation would endure.

There was only one nagging doubt left in my own mind about the efficacy of this particular *Incwala*. One of the final acts of the whole ceremony was supposed to be rain making. If rain did not fall on this concluding day, then the nation may be facing misfortune, or worse. The reason, as I learnt later, was not really rain making at all, but rather to do with the rites of final purification. On the morning of this last day, all the accoutrements of the whole week's cultic offices were gathered, together with the gall bladder of a black bull which had been ritually pummelled to death by the young warriors, and sacrificed on an immense funeral pyre. It was essential during the course of the last afternoon that the flames of this fire are quenched by new rains. I hated to be the bringer of ill tidings, but when we had started out from the farm in our dilapidated grey Land Rover there was not a cloud in the sky, which may not have boded well for the Nation's future. However, absorbed as we had been by these remarkable rituals, we had failed to notice a cloud, somewhat larger than a man's hand, looming over Sheba's Breasts and making its way towards Lobamba, spiritual heart of the Nation. The shadow of this fast-growing thunderhead sped across the ground and put out the sun. As we looked up and saw the sky grow dark, then black, we also realised that we were about to be engulfed in another cataclysmic tropical storm.

We raced back to the Land Rover, which as fate would have it was open topped, you may recall. Paulo put the key in the ignition and the jeep emitted a paroxysm of smoke and coughed into life. We rattled over the rough ground, regardless of engineering deficiencies, and weaved our way back onto the main road. By that time the leading thunderhead had become a monstrous column with a roiling summit like an atomic explosion. Inside this cloudy pillar vertical streaks of orange and green lightening flashed and cracked, turning it into an ominous and incandescent Roman candle. With one ultimate detonation, which echoed around the ancient mountains of the Mdzimba, the heavens opened and after a few heavy spots the thunderstorm cascaded over us like wild, breaking surf. There was a yellow tarpaulin in the back which we tried to rig as a canopy, holding the leading edge against the windscreen while the windscreen wipers oscillated aimlessly in the empty air on the top of the bonnet, to where they had been bent years before. We realized as the vehicle reeled distressingly across the road that trying to stay dry was futile, and abandoning the tarpaulin we succumbed to our fate. We arrived back at the farm some time afterwards looking and feeling like sodden cardboard.

I read some years later in an account relating to *Tincwala* of the past that no matter how heavy the storm on the ultimate day, the people are so thrilled with the rains that they do not seek shelter, but rather, drenched to the bone, they round off the performance with a national chant. We had done the 'drenched to the bone bit', and our chant was for hot showers and stiff gins and tonics all round. I don't know what medicines had been used in Lobamba that week, but they must have been incredibly potent to induce a storm of this magnitude. As I sipped my gin and tonic and reviewed the day, I realised to my astonishment that I had just been introduced to one of the last surviving African monarchies of this great continent. One European power after another, be it the British, the French, the Dutch, the Belgians or the Portuguese, had systematically and maliciously dismantled nearly every other one, from Ashanti to Zanzibar, in their colonial scramble for Africa. But Swaziland was small enough and withdrawn enough that it had escaped this regal emasculation. It had retained its extremely rich culture and history, which now began to hold just as much a fascination for me as the new revelations of the Stone Age. If it were possible, I was even more entranced by Swaziland and its peoples, past and present, than before.

* * * *

We were talking about the events of the day that evening in the chintz drawing room, about the *Incwala*, about the king and the traditions of the Swazi people. We were by now mellowed with a sufficient dosage of the 'Queen's Tears,' as the Zulus used to call gin in the old days, an allusion to Queen Victoria. The storm was beginning to abate outside, with only the occasional *doner und blitzen*. I declared that I was fascinated by the power that witch doctors and traditional healers could wield over the indigenous people, even over the King. My father-in-law, Farmer Joe, agreed, and then recalled an event that certainly put things into perspective for me.

Many years before, he related, when he had first arrived in Swaziland, he had suffered a lightning strike on the farm. It turns out that this part of

Africa experiences more electric storms than just about anywhere else on earth; from my limited knowledge so far I could certainly attest to that. Most storms passed off benignly, but in this case a lightning bolt struck a hut where some of the farm labourers were sheltering, and it killed two women stone dead. The other workers fled uncontrollably into the veld, abandoning the bodies of the unfortunate victims where they had fallen. As far as they were concerned this was *butsakatsi*, witchcraft. In fact, the fields were bewitched, along with the whole farm. A *'tokoloshi'*, a malevolent trickster spirit much feared in southern African domestic mythology, had descended upon it. No one would work on the land and no one would touch the departed.

Next day, the labourers did not return to the fields and the crops sat unreaped. The farm was deserted. The following day, and the day after that, the same thing happened. No workers came.

"So what was I to do?" said Joe, despairingly.

His long experience of farming with the fellahin in Egypt had not prepared him for this. Superstition was ruining his monthly figures, which were not so good at the best of times. He took advice from a variety of *'bafundzisi.'* Everyone told him he must employ a *sangoma*, a traditional Swazi spirit healer, to get rid of the evil spectre hanging over the fields. He was unimpressed. He felt he didn't need some half-naked African witchdoctor to sort out a simple labour dispute. But the problem continued. Days went by, and still no work on the farm. Eventually he conceded, and with great scepticism he hired a *sangoma*.

"And if you're going to get one, get a good one; get the king's *sangoma* for a proper job," he was advised.

Duly the next day the king's *sangoma* appeared on the stoep of the farm. Farmer Joe, who had been brought up with Edwardian discipline in rural Sussex, was not amused. Before him sat a noisome individual dressed in skins, wearing many rows of multi-coloured beads, bones and bangles. His hair was matted in dread-locks glistening with dark-red ochre. He demanded a sum of ten pounds for his services, which in those days was an outrageous amount. But Joe was forced to agree and the doctoring began. The *sangoma* ambled around the whole boundary of the farm, muttering incantations and spells. With great solemnity he appeared to puke at each of the survey beacons that defined the shape of the plough land. Presumably he was expectorating powerful herbal potions which would drive away the evil. After an hour or so, he declared the *'tokoloshi'* gone. The exorcism was over.

At once, the labourers returned and work resumed. Joe finished the story, and with a twinkle in his eye he said:

"And do you know, whenever there was an electric storm in the valley, lightning struck all around Malkerns but never struck our farm again. Ten pounds well spent, I always thought."

Joe was a fund of stories about pioneer farming in the old days, and I was a very willing listener. The present storm had reminded him about the hail rockets they used to use years before, actual rockets fired at the storm clouds to try to break up the hail before it broke them. Joe and the farm had suffered badly from hailstones in the fifties. One year, for example, he and his wife, now my esteemed mother-in-law, had been sitting in the evening on this very stoep with most of the farm, a huge acreage, planted with the best cotton they had ever grown. The bolls had reached that critical moment when they had just burst, revealing their soft downy tassels of white, like balls of wool hanging from the bushes. Tomorrow they would start to pick. Incidentally, when they do pick cotton on the farm, as I later saw, the crop is all reaped by hand and to the literary eye it does somewhat resemble a scene from *Uncle Tom's Cabin*. They were already counting their rewards, money that would allow them to begin to build their dream house to replace the mud and corrugated iron. They had the plans in a drawer ready.

Just when they were into their second warm gin and water, a colonial habit which fortunately I have never acquired, a storm crashed over them with hailstones the size of tennis balls - they showed me the dents all over the caravan roof - and within minutes the whole the crop was on the ground, ruined. No house, no third gin and water, and it was 'back to the bank manager' time. Thereafter they tried using hail rockets. They would erect this rickety tripod on the lawn, point the rocket at the growing blackness and pull the lanyard. Whoosh! Amongst other things it must have been great fun, and it did seem to work, until the fateful day that one went astray and, arching round, took off horizontally in a direction opposite to that intended. Sparks showering from its tail, it crossed the road and flew straight for the house of Brocklehurst, the family doctor and a bosom friend. As the mild-mannered medic was just emerging from his front door, this fiery apparition was making a beeline for him. Thanks to his agility, it narrowly missed him and the house and exploded with an almighty bang in his back garden. He was not amused. Diplomatic relations were temporarily suspended until Joe could convince him there would be no more rockets. Joe was reluctant to grow cotton ever again, hence the hail-resistant pineapples.

CHAPTER SIX

MAKING A PLAN

So geographers in Afric maps
With savage pictures fill their gaps
And o'er uninhabitable downs
Place elephants for want of towns.

Jonathan Swift

The old year came to an end and it was time to consider our next adventure into the unknown, in this case into the low altitude Swaziland bush (the bushveld). After the routine pantomime with the telephone and a great deal of hand gesturing which clearly made no impact at all on the Malkerns operator since she was two miles away down the road, the drainpipe line was connected to Johnny in 'The Place of Winds.' With harmonised, staccato bleating and with an enriching electro-static accompaniment we made a plan to meet down in the lowveld on New Year's Day. 'Make a plan' was the local jargon for cobbling something together, as in 'Och Man, we'll make a plan, eh?' For this plan to work, however, we would have to re-habilitate 'Old Blue' of Komati fame, as we would need to travel in our own 4x4, and I didn't even want to contemplate the engineeringly senile 'Grey Heap.' A mechanic was duly summoned and for hours I anxiously watched his emergent bum in greasy overalls cork-screwing around the engine compartment. He too was making a plan. On the appointed day, the Land Rover was ready. Well, if not ready, at least in a 'go' mode.

Remember the earlier continental comparison between Africa's profile and Livingstone's inverted soup plate? We had already been to the highest point of the upturned base, the highveld at Ngwenya. We had careered down the rolling hills of the convex curve of the middleveld and seen, amongst other places, the Ngwempisi River. We were now working our way towards the outside edge, the flat bit, the flange where you put the prune stones if like me you eat stewed prunes from a soup plate, always remembering that we are talking about the underside of the flange. Anyway, we were now driving eastwards towards the sea. I just hoped that when we got to the edge 'Old Blue' would recognise the final frontier in time to stop, otherwise we might truly be in the soup.

The old banger was surprisingly well-behaved that day and positively purred down from Malkerns towards Swaziland's second largest town, Manzini,

'The Place of Waters.' Mind you, by definition we were going downhill all the way, physically if not metaphorically. A notice greeted us at the outskirts - 'Manzini: Welcome to the Hub'; the town sits in the centre of a vaguely circular country. It was an agricultural settlement, replete with post-office, several general stores, shops selling lurid instruments that could be used to effect a gender change in cattle, and a Portuguese restaurant called '*The Mozambique.*' 'Piri-piri prawns a speciality,' read the hoarding, with a picture of a monster if deformed shrimp smirking out of an arbour of lettuce and chilli peppers.

Eschewing the lure of the George Hotel at the far end of the town – 'lunches, dinners and ladies bar' – we drove on out of Manzini through a delightful avenue of jacaranda trees, some still sporting striking sprays of lavender-blue flowers. Actually, the species originally comes from Brazil, so in our growing purist appreciation of Africa we didn't really look at them. The road continued down past the *'Gum Tree Bottle Store,'* named after a nearby stand of eucalyptus, imported from Australia, so we didn't really look at those either. That was the last outpost of the modern world. Beyond, the contours softened, ever downwards, and on either side of the road there appeared short scrubby trees, becoming ever denser. After half an hour, breasting a low rise, the landscape suddenly opened in front of us and there was the lowveld, spread out like an immense level rift thirty miles wide and stretching to the right and left as far as the eye could see, a vast featureless plain covered, from what we could make out, with a uniform mantle of olive-green thorn trees. As we stopped the Land Rover, a little bare-footed herd-boy clothed in an unexpected Captain Whizzo t-shirt and khaki shorts, both substantially holed for better ventilation, shuffled past us with a goad, driving a very large black and white bull many times his own size. The lad kept craning his head round to watch us as he walked into the bushes after the bull. Where he and the bull were going was difficult to tell for the press of trees.

Far away in the distance the soup plate analogy came to an end, literally. Defining the farthest edge of the lowveld trough was a continuous line of low hills with an almost exact level crest. It was a dark smudge along the horizon, at first almost indistinguishable from a thin bank of low cloud but now perceptible enough to connect with the ground. This was the face of the Lubombo mountains, a two thousand foot high escarpment running for four hundred miles from the Limpopo river in the north down to Maputaland and Zululand in the south, and this is what cut Swaziland off from the coast. Beyond the scarp face of the Lubombo the slopes slowly dissipated their newly gained height down to the Rio Maputo and the Indian Ocean. But that was in Mozambique. Swaziland was land-locked. The only ocean it owned to was a marine vista of thorn trees. Coaxing 'Old Blue' back to life, we launched ourselves down onto this acacia sea to take a closer look.

We had finally arrived at the real African savannah, a region of grassland with trees. This thorn-scrub stretches in an unbroken line down the eastern side of the African continent, from Ethiopia through Kenya, Tanzania, Malawi, Zimbabwe, and Botswana all the way to the Eastern Cape - from the Horn to South Africa - getting on for four thousand miles of uninterrupted 'bush'. It's what the old diehard colonial surveyors would mark on their maps MBA – Miles of Bloody Africa!

But this is also TV Africa, with its stereotypical flat-topped thorn trees and open grassland. It's an ecosystem that has given rise to the greatest density and diversity of large animals in the world. That's why there are so many wild life series about Africa, migrations on the Serengeti plains and all that. But why? It's the rainfall. The bushveld in Swaziland receives about five hundred millimetres of rain a year, on average, though there is no such thing in these parts as 'average' rainfall. It's either sheeting with rain or totally parched. There are summer rains that fall at the hottest time of the year either side of Christmas when the sun is blazing overhead and the thermometer at this low altitude, near sea level, can climb to forty degrees Celsius by breakfast time. When it is hot like that the rain begins to evaporate almost before it hits the ground. This means that the plants which grow here have to be adapted to these conditions. There is intense competition for moisture, hence the huge variety of trees that live here, hundreds of different species just in this part of Africa alone. They are generally all shallow-rooted, spreading themselves wide to catch any wetness they can, spaced apart among the grassland, looking in some ways like a deliberately planted home-counties park. In fact, this environment is sometimes called 'parkland savannah'. Who knows, maybe that's where Capability Brown got the idea from?

Evaporation sucks all the soluble minerals - the calcium, potassium, sodium and the like - all the minerals which sustain life - to the surface, or near it at least, so that the grasses and trees which live on these soils are highly nutritious for everything from aardvarks to zebras. Here in the bushveld these thorny trees and swaying grasses are so packed with mineral supplements that they can support a huge variety of browsing and grazing animals, ones that eat leaves such as giraffes and others that eat grasses like buffalos. The downside is that there are many months in the winter without any rain, maybe up to nine months or more at a stretch, so everyone has to be able to go without a drink for long periods to survive. Thinking of my own personal circumstances that would be a tall order.

A wider variety of animals and greater herds of them have evolved in the African savannah than anywhere else on Earth. Significantly, this includes our own genus, *Homo;* it's from Africa's savannah grassland that all our predecessors came, going back millions of years. We didn't come from the Welsh Marches after all, or even from Sussex and the English Home Counties, as the Piltdown forgery had led us to believe, but we had our origins in Africa, every single one of us. Considering the immensity of these ideas, I was most unusually rendered speechless.

* * * *

We were now well into the bush and nearing the point on the map where, according to my navigating, we were to meet up with Johnny. "Meet me under such and such an acacia," he had said. Was he deranged? There were thousands of acacias, all looking the same. But we found him and it was a relief to connect with him in this unfamiliar landscape. Wishing him the compliments of the New Year, I told him enthusiastically of our experiences, of the *incwala*, of *sangomas*, the *Ngwenyama* and now the bush.

"Yes, yes!" he said encouragingly. "You're getting the hang of it. But let me introduce you. This is Ralph." Beside him in the passenger seat of his Land Rover sat someone my own age, dressed in khaki shorts and an epauletted khaki shirt. He was soft-spoken, with a slightly boyish face and blinking eyes.

There was something of the Billy Bunter about Ralph, an impression assisted no doubt by his round wire-rimmed spectacles and similar physique, round, I mean. I wondered how he fitted into the scheme of things. He knew of Sue, Farmer Joe's daughter, because he too had been brought up in Swaziland. What I didn't yet know was that Ralph was an encyclopaedic expert on the bush, and that he was to become one of my closest friends.

We drove off in tandem, eating their dust. I was afraid to hang back in case we lost them. Abruptly they turned off the dirt road onto a bush track, which finally led after some convolutions to a high, wire gate. A smartly dressed Swazi in a quasi-military uniform - moss green sweater, pressed shorts, knee-high khaki socks and polished shoes - saluted to attention. We were now entering Hlane Game Sanctuary, private hunting preserve of the king. Ralph was the king's senior ranger. After a few more turns, we arrived at a mud-coloured thatched house deep in the bush surrounded by tall thorn trees where Ralph lived. A small grassy patch had been cleared in front of it on which a group of impala were grazing, and from behind the branches of a gnarled acacia chewing with great deliberateness a giraffe peered haughtily down at us with long effeminate eyelashes.

We stopped in a cloud of dust and went inside. The concrete-floored main room was sparsely furnished with old leather armchairs and an ancient sagging couch, a zebra skin thrown carelessly over it. A few bottles with guttered candles sat on a Victorian sideboard at the back, and a rustic coffee table of dark, knotty planking held a couple of ashtrays made from upside down Land Rover piston heads. One side of the room was open to the lawn and the bush, except for a knee-high wall to keep inquisitive animals at bay. "Coffee?" asked Ralph. He motioned to his Swazi batman Philippe lurking in the shadows and spoke quietly to him in siSwati.

There were birds everywhere. On the stone-slab bird table outside there where flocks of flashing, yellow-rumped, black-eyed bulbuls, with their constant 'chit, chit, chit' chatter. Above, in the tops of the thorn trees, a number of silvery-grey louries screeched every few seconds. This was the famous 'go away' bird, named on account of its call, a hoarse 'Go weeey.' Across the grass, hanging from a leafless tree, was a large colony of masked weavers. The males, a deep canary yellow with black eye-patch, were completing the construction of their individually built pendulous nests woven from grass blades. They had been crafted with a narrow downward-facing tunnel for access and to prevent boomslangs - green tree snakes - from attacking their young. They each fluttered at the entrance to their new creation, hoping to entice the less colourful females to join them.

Most striking of all were the sunbirds. Outside the front wall of the sitting room was a row of aloes, thick, spotted succulent leaves spreading across the ground, out of which grew tall slender stems topped with clusters of small shell-pink bell-like flowers. Among these darted tiny birds like humming birds, whirring their wings to flit from one flower to the next, dipping their long curved beaks into each bell to drink the nectar. There were several different species, malachite sunbirds and double collared sunbirds among them, all with iridescent green or black or crimson plumage. I was fascinated by their speed and agility. The impalas looked up at a movement. The coffee had arrived.

"David here is interested in our archaeology," said Johnny innocently, "so I thought we would show him around."

We talked at some length about the work I had conducted researching prehistory and ancient climates in the Near East, and as we did so, the corm of an idea was planted deep in my imagination. I described how I had spent the last five years in the Gaza Strip looking at Stone Age man against the backdrop of climatic change during the Pleistocene Epoch, the name geologists have given to the last geological period before the present. Gaza was an explosive place even then and I had been variously bombed and shot at during my time there, which tended to concentrate the attention somewhat.

The Pleistocene epoch covered the last two million years of Earth history, save for the last ten thousand, and what was really significant was that it was during this time that humans emerged onto the world stage. The Pleistocene was also the time of the great Ice Ages. When I was a lad in short trousers, living atop that moraine in Shropshire, we were taught that there had been four major ice advances during the Pleistocene, following the analysis of two German geologists in 1909. The newest evidence I spoke of that morning at Ralph's bush house suggested there had been not four but at least twenty of these catastrophic climatic upheavals during the same time.

"So what happened in Africa during these times?" came Ralph's deceptively simple but crucial question. "Isn't this where we all came from? What happened here?"

"I don't know," I replied lamely.

"Well come along, Price Williams," said Ralph again, "we'd better find out, hadn't we? No sense in buggering about in the U.K. in the snow, or the Middle East in the sand. Africa's the place to be, man. This is where it all happened. Let me show you."

And with that we got up and drove deeper into the bush.

* * * *

I had never really paid much attention to African wildlife before that day, let alone ever been on safari. In Britain, the nearest I had approached to it was a couple of TV series during the late 1950s. One was called, appropriately enough, 'On Safari.' This was a bizarre, black and white concoction created by a Belgian film maker Armand Denis and his Anglo-Russian wife Michaela. The programme used to open with jungle drums *à la* Tarzan of the Apes. Armand, who looked like a cross between Alan Whicker and Mr Pastry, would drone on in heavy guttural tones about some great philosophical gibberish he had thought up on his last trip to the wild. His wife Michaela was a blonde, glamorous piece who always wore loads of make-up, even in 'N-Goro N-Goro Crater,' a place she talked about a lot. I had a lot of time for Michaela. Ooo, those open bush shirts. But the dialogue was somewhat stilted. Picture the scene. The lovely Michaela is cooking over an open fire. As Armand films he ponderously intones: "The lovely Michaela is cooking over an open fire." The whole thing was rather naïve, but I do remember the occasional animal in the distance.

The other series was 'Zoo Quest,' with David Attenborough. Here was the young David holding a ring tailed lemur or some other obscure creature on his shoulder. He would flutter his eyelids and speak lyrically to camera about the beauties of nature whist this turdulent animal heaved its bowels and defecated all over his jacket; not very illuminating.

Sans Michaela, we drove into the bush. Hlane was a huge area of a very special kind of wilderness. Ralph mentioned that there were at least twenty different species of acacias in the reserve, the dominant one of which was the knobthorn, *Acacia nigrescens*. These were tall trees more than thirty feet high, with deeply fissured bark covered with large conspicuous knobs, each carrying an evil-looking hooked thorn. He explained that the tree had sweet resin beneath the bark, much loved by elephants, and the thorny knobs were an attempt by the tree to stop elephants stripping it bare. Another acacia was the umbrella thorn, *Acacia tortilis*. This is the one so often seen in wildlife films, with its high, flat crown. It carried both hooked and straight thorns and it grew high like that to reduce browsing pressure from smaller animals. Somewhere in the bushveld was even a small Swazi acacia, *Acacia swazica*, which grew only here and nowhere else on earth. Compare that, he said, to the Nile acacia, *Acacia nilotica*. The name spoke for itself. It grew from Hlane all the way to North Africa. He pointed to another especially fine looking tree:

"This tree is the Transvaal silver leaf which is a sure-fired indicator of sandy soils. And this is the marula. The locals brew a potent beer from the fruit."

He knew every plant that grew in Hlane. Every tree had its place and its own way of coping with pressure from browsing animals like kudu, and its own way of dealing with drought. During the long dry winters some trees were capable of breaking off up to ninety per cent of their external limbs to reduce their water requirements.

Ralph pointed to the various grasses. The reddish tinged grass was *rooigras*, especially nutritious for grazers. The white headed grass was an indicator of disturbed ground. This tree grew only on rocky outcrops. That one only grew where there was plenty of underground moisture. The list was endless and impressive. Ralph was building up for us a fingerprint of each microenvironment, a profile of the infinitely varying habits of the bushveld and we hadn't even seen any animals yet. There was so much more to the bush than game viewing.

But there were animals too. At that moment a small herd of zebra trotted across the path in front of us tossing their heads.

"These are Burchell's zebra. Notice the shadow stripes."

Between the darker stripes were paler ones that mirrored them. The stripes must be for camouflage, I thought.

"The stripes are not really for camouflage. You can see them too easily. They are actually to confuse predators, lion for instance. When the herd runs together all the lion can see is a mass of flashing stripes, like a strobing light. It finds it difficult at speed to focus on any one individual animal and so gives up the chase; well, that's the theory anyway."

I didn't know that. In fact, I didn't know this language at all. There was a whole grammar and syntax of the bush I had never come across before in my life. I was determined to learn it.

The next herd we saw was of wildebeest, archaic-looking animals that loped along with heavy head and swirling tail. They had a curious listing movement of the front legs and shoulders, like a sort of shrug. I remembered that my father-in-law Farmer Joe's Swazi name was '*ingongoni*,' 'wildebeest.' I could see why now. He'd had a bad fall playing polo years before and walked with a slight rocking of his shoulders where his back has fused. We watched as the wildebeest with their new-born calves shambled off.

"Makes very good biryani, does wildebeest fillet," interjected Johnny with a wry smile. That seemed unnecessarily callous until I realised that humans had changed from being largely vegetarians to become meat-eaters and predators in the African savannah. How long ago had that happened, I wondered? The corm of the idea, which had been planted earlier that morning at Ralph's house, began to sprout.

Rounding a gravelly corner Ralph stopped the Land Rover and pointed between the thickly growing thorn trees at what seemed at first to be two granite boulders. They turned out to be two rhinos and they were coming our way. These were white rhinos, Ralph said, Africa's third largest land mammal, after the elephant and hippo. This pair was a mother and adolescent male. The mother alone weighed a massive 1600 kilograms.

"They are not white at all, but wide-lipped. They are grazers. The black rhino is the browser. See, you can tell by the dung."

He picked up a piece of dried dung from the track by the side of the Land Rover and handed it to me. I looked at it intelligently, or as intelligently as one can look at a large tubular lump of ordure. It looked like dry, compacted grass clippings.

"Ah, yes," I muttered sagely.

The rhinos were advancing. You could see the mud on their flanks and the female's mean-looking horn. They shambled closer still. That was some long, slender spike she had on the end of her nose. I had visions of her skewering her horn right up Ralph's fundament if he didn't get back in the vehicle.

"These rhino have poor eyesight but a very good sense of smell. Don't worry, they don't charge, normally."

Let's hope these were feeling normal, that's all I had to say. But as we watched these gentle giants, I remembered that species of rhinoceros have been around a long time. They have even been found in Europe as early as two million years ago, at the opening of the Pleistocene, but now of course there were none left there. It seems that even here in Africa only two species remained today, the white and the black, and these had been hunted almost to extinction, not for their meat but for their horn. In China powdered rhino horn is regarded as an aphrodisiac and in Yemen they made handles for their curved daggers with it, which for them is a huge male status symbol. A massive conservation effort would be needed if these animals were going to survive even another few decades. It was very poignant, watching the last of these prehistoric behemoths as they shambled away into the thorns.

We picked up Old Blue and headed in tandem again, east towards the Lubombo, driving through beautiful knobthorn woodland, sending warthogs scurrying for cover, tails up like car radio aerials. Kudu and bushbuck stood in the shadows of the far-off glades. Way in the distance more rhino could be seen lumbering through the bush. The sunlight glinted through the overarching branches of the trees. It was all so lovely and so primordial, as if this is how it had always been. We reached a ford across the Mlawula River. Cape bull rushes grew by the water's edge, and down-stream two waterbuck stood knee-deep in the river grazing on the reeds. We left Hlane and crossing over a small ridge came into a hidden valley that ran parallel to the Lubombo Escarpment which now towered above us. This was the Nkumbane Valley, exquisitely unspoilt, pristine African savannah.

A shower of impala bolted out of a patch of long grass and leapt almost over our heads, their tan and cream flanks shining with health. When they had bounced far enough away they slowed to a trot and merged into the trees. Ralph stopped his Land Rover. We stopped behind and got out.

"This is not part of the Sanctuary, but we are hoping to declare it a game reserve soon. It will be called Mlawula," Johnny told us.

He pointed to the ground and picked up some tiny slivers of what looked at first sight like coloured glass. He held them out in the palm of his hand. They were microliths. I had never seen any of these miniature tools before, but I knew what they were. They had been made by Late Stone Age people hundreds, if not thousands, of years ago, tiny blades and flakes no bigger than half a match crafted from the finest rocks, with colours like vivid orange, transparent green and milky white. The little blades were made from carnelian, heliotrope, agate and other semi-precious stones. They had been manufactured from solidified gas bubble geodes, like the agate ones polished up and exhibited in gem shop windows. The lavas in which these geodes were found and which made up this Lubombo range exploded out of the Earth's crust when the southern super continent of Gondwana was breaking up two hundred million years ago. It seems these cliffs above us held a key to how Africa as a continent came into being. And these tiny fragments would tell the story of the latest of the Stone Age peoples in Africa. With this information the whole cycle of human existence in Swaziland had come around full circle. The sprout of that morning's idea at Hlane put out branches and began to bud.

"And what about these?" asked Johnny, holding out some coarsely-made chipped flakes, each about the size of a credit card, but skewed in a more rhomboidal shape. "They look a bit like a Father Christmas boot," he added, with a somewhat obscure if seasonal twist.

At that time I had no notion how incredibly prescient these seemingly casual moments in Mlawula were to prove, but in retrospect these shapes changed my life.

We continued up a disused track and forked into a narrow valley beneath the rock face of the Lubombo. The scenery was stunning. It was all polychrome and verdant, shining in the mid-day heat, the long grass, luxuriant foliage, and yellow and white acacia blossoms among rich green filigree and dark gnarled trunks all melding into a kaleidoscope of unsullied abundance. The track wound on into what appeared to be a dead end. In order to get out of the valley we would either have to turn around or climb the face of the escarpment, which was ridiculous. I quickly learnt that we were taking the second option. We were going to climb the face of the escarpment.

"Can you drive that thing, Price Williams?" barked Ralph, leaning his head out of his side window. "Low ratio!" he added.

The track suddenly raked up to what felt like forty five degrees. Ralph and Johnny went first, wheels spinning and rocks scattering in cascades. I lost sight of them in the dust. It was our turn next. I banged the levers into low ratio first gear and gingerly pressed the accelerator. 'Old Blue' leapt forward onto the slope and, attacking it with a startling vigour, began to climb. It suddenly felt as though the track had just been transformed into a gigantic demonic cake walk. The vehicle bucked and shook like a rodeo steer and we were thrown about against the sides and the dashboard like a cowboy at his first attempt. The front wheels were thrashing uncontrollably from side to side, ripping the steering wheel out of my hands and spinning it violently left then right. We were enveloped in clouds of dust and could hear rocks underneath thumping against the chassis and pinging against the exhaust pipe. In a temporary break in the dust clouds, I saw Ralph pointing out of the window of the vehicle in front at a delicate paper-barked bush with flowers like little golden balls. "Acacia swazica," he bellowed. Christ, how had he had time to notice that at a time like this? What kind of man was he? I later looked up *Acacia swazica* in a reference book on acacias. 'Inhabits remote rocky slopes,' the authoress had written. Was she ever right!

We fought gallantly onwards and gradually, slowly but surely, we reached the heights, the top of the Lubombo. My wrists ached, my shoulders were bruised and we were covered in dust, but we had done it. 'Old Blue' had come through triumphant. Drained, we pulled up under a broad-leaved rock fig and got out. The view was staggering. The whole of the bushveld lay at our feet, and looking west into the interior, way in the distance, ridge behind ridge, range behind range, the whole of the great African Escarpment unfolded before us. It was magnificent, a moment I shall never forget. It was marvellously heady, like the first draught of a fine wine. At that precise moment what should appear from Johnny's hamper but a couple of cold bottles of Neethlingshof Cuvée Blanche, four crystal glasses and a polythene box full of Stella's excellent meat pies. I took a glass of wine, matt with condensation from the cool vintage, and sat down on a rock some distance away, drinking in the scene and the wine by turns. I stared at the far horizon.

Somewhere among those blue ridges lay the Ngwempisi river and the hand-axes of the Old Stone Age. Somewhere there was Ngwenya, the Middle Stone Age and the Komati rock paintings with the shamans in trace. And somewhere beneath us in the bush the Late Stone Age people had made their carnelian micro-blades. What a country. This was the Swaziland of the Ngwenyama Sobhuza, traditional ruler of this ancient African kingdom. How well his ancestors had chosen for him. And for me, the whole of human time was here. I turned it over in my mind - highveld, middleveld, bushveld, Lubombo – four totally different ecological zones, four different environments, with different rainfall, different altitudes, different temperatures, different vegetation, different human opportunities, all throughout human existence. The buds of my Hlane idea opened up into full leaf as I watched.

I silently reasoned that if there had been climatic change here in the past, it was bound to be reflected in the shifting patterns of these different ecosystems, and in turn that would be mirrored by differences in the human response, that is, the archaeology, all the way back to the beginning of human time. And here in Africa we still had elephant, rhino and antelope galore. To my astonishment I realised that unlike Europe or America, which had lost all its large mammals at the end of the Pleistocene, Africa was effectively still in the Pleistocene. What an opportunity for studying man in the Pleistocene, by actually being there. As I sat on that rock at the crest of the Lubombo that New Year's Day, the enormity of the concept dawned on me. I turned it round in my mental vision, looking at it this way and that, testing for flaws, then seeing with great clarity the implications of what I had just conceived.

I walked back to the Land Rovers and helped myself to an overflowing glass of the ice cold Cuvee. Taking a gulp I launched into a rapid, if garbled, description of the project I had in mind. I pointed to the western sky and began,

"You see, if you put this with that, and that with this, and look at it this way, and do it that way, and bring in archaeo-botanists, some geomorphologists, and a few paleo-zoologists, and dating experts, and a team of prehistorians, and a paleo-ecologist or two, we could crack the code of what happens in Africa during a globally cold period, in fact, through the whole of human time. Swaziland would be the ideal field laboratory. Do you see? Sorry, I probably haven't made myself very clear."

"No, David," said Johnny, "you've made yourself abundantly clear. That's what I was hoping you would say."

And at that moment SARA came into being, the Swaziland Archaeological Research Association, which would absorb all my time for the next fifteen years and take me to every corner of this beautiful kingdom and far beyond. Having unwrapped my gift of Africa I had now just made it work.

"Well, come on Price Williams," said Ralph. "You'd better get on with it then."

So we did. And that is how I ended up in the House of Lords!

CHAPTER SEVEN

DR WATSON'S PATENT COLLUVIUM

The House of Lords is like a glass of champagne that has stood for five days.
                                            Clement Attlee

I had never wanted to go to Africa. Now I was there, I never wanted to leave. The three weeks I had spent in Swaziland had brought about such a fundamental revolution in my thinking and life-style, it was impossible to absorb the sheer scale of it. As the last hours of our visit ticked away, I found myself more and more despondent at having to return to Britain. The family threw a small party on the night before we left. It began as a fairly simple affair, not like the boisterous atmosphere of the lamb roast. We talked together in muted tones, as though there was someone gravely ill in the house. Sue had not visited her old home for some years, and my parents-in-law must have felt it would be another few years before she would return, especially as we were expecting a baby. I had other ideas. I had been buoyed up with so much exhilaration for the new expedition we had discussed that New Year's Day on top of the Lubombo, looking across the bushveld. Without really thinking about any logistical problems and not having talked it over with Sue either I was already planning how we would return.

Farmer Joe, toughened by half a century battling with agriculture in the sub-tropics, nevertheless had his softer side too. He was very conscious of 'famly' as he called it. He still spoke with a very slight Sussex down-land burr, and having fortified himself with a couple of stiff Bell's whiskies, known locally as a ding and a dong, he rose to his feet to give his 'State of the Dynasty' address. At the dinner table that night he had his two sons and one of his daughters, plus daughter-in-law, a couple of grandchildren and me. As he ranged over the family's career he ended on a further welcome to me, as the most recent addition, greeting me into the inner circle. Then he raised his glass and pointing it in my direction, gave a toast:

"I looks at you, and I shudders!"

The party then descended into a more animated affair, and I half expected Joe to launch into his earthy rendition of 'Susanna's a comical pig!' and other bucolic anthems from rural Edwardian England, with full farm yard accompaniment. But time was running out, and we had a plane to catch the next day.

After tearful goodbyes we drove away from the farm, from the mud house with a tin roof that I had come to love, from the stoep where my conversion had taken place, from the garden with its bougainvillea and cannas, down the now familiar farm road and away up the Malagwane Hill passed Sheba's Breasts. Retracing our journey of three weeks ago, this time with a totally changed view of life, we passed through the border post, crossed the Transvaal and reached Johannesburg. With so much going round in my head I hardly noticed the distance. But I knew with a powerful and absolute certainty that it would not be long before I would be back, we would be back, in fact, all three of us would be back!

* * * *

In England I fought the good fight on behalf of my new obsession, struggling to develop a base from which to operate and raise funds for our proposed research. To this end, one of the top mandarins of the University with which I was loosely affiliated had recently been elevated to the peerage after a life of self-less service to matters colonial. In fact, he had strong ties with southern Africa and knew Swaziland well, having been a minor political player there in the era after the 'winds of change' phase of Commonwealth evolution. His solution was to put together a committee of like-minded philanthropists to help us out, a committee that he himself was quite happy to chair. I was truly flattered, though I wasn't quite sure where it was leading or what I would do with this new committee, particularly since the members all seemed to be poor as church mice. For example, one household name had made his reputation climbing in the Himalayas or the Andes or some Alp or another. I couldn't quite see how mountaineering could be woven into the fabric of our proposed expedition into African prehistory. You could drive to the highest point in Swaziland with a Land Rover. I had already done it, without crampons or oxygen.

Slowly the grandiose Swaziland Archaeological Research Association Steering Committee took shape. It was composed of two peers of the realm, two MP's with African interests, a number of Foreign and Commonwealth supernumeraries, miscellaneous academics with high-sounding titles, a tame lawyer who listed archaeology as his hobby, and me. Our first meeting was to be at the House of Lords, no less. I was moving in elevated circles. Actually it turned out to be ever decreasing circles, but more of that shortly. I contrived to manufacture various papers for the meeting, carefully disguising the fact that so far I knew very little about Swaziland or its archaeology. However, I banked on the principle that I was likely to know more than the rest of the committee, by whatever small a margin, and that in the country of the blind the one-eyed man is king.

On the appointed day I made my way to St. Stephen's Gate, the equivalent of the tradesman's entrance of the House of Lords. I wore my best Montagu Burton suit, bought in last year's sale, and a pair of shiny black shoes. I had wanted to add a tie with giraffes on it, to give the whole ensemble a certain environmental verisimilitude, but my wife prevailed and I stuck to the plain blue. I treated myself to a taxi to the entrance. It was a marvellous moment, saying to the cabby, 'House of Lords please!' Driving down St James's I mused on the difference between travelling to the Palace of Westminster in a taxi, wearing a pinstriped suit and black brogues, versus driving through the African bush in a clapped-out Land Rover wearing a pair of dusty shorts and sandals.

I entered through the gate under the great tower as Big Ben struck the hour. A policeman hailed me by saying, "I don't recognize you sir." I agreed with him, saying that it was probably on account of the suit I was wearing. I confided that I had hardly recognised myself that morning when I looked in the mirror. When he repeated the statement, I realised my mistake. This was his formal way of telling me I was a 'stranger'. It was a challenge. Again he enquired just where did I think I was going, or something to that effect. I drew myself up to my full height and said:

"I am here to attend the inaugural meeting of the Swaziland Archaeological Research Association."

He stiffened visibly and with a great deal more respect said:

"Certainly Sir, Committee Room 18 on the first floor."

I was deeply impressed. I had come a long way from Hlane and the Mlawula Valley to be here today. Imagine what Johnny or Ralph would think if they could see me now, taking our case all the way to the House of Lords, so to speak. Well, not quite that, but the phrase had a certain ring to it. We had dreamed up that title, SARA, over a couple of bottles of ice cold Neethlingshof Cuvée Blanche under a broad-leaved rock fig on the rim of Africa. I suppose strictly speaking that had been the inaugural meeting. Anyway, now here was SARA being recognised at the House of Lords. What would the Stone Age microlith makers of the Nkumbane Valley make of it all, six thousand miles away in the beauty of the bush? Well, they probably wouldn't have been interested in the slightest, and quite right too. As I climbed the carpeted staircase I wished nervously that I could be back there, squinting through the knobthorns at the rhinos. After all, I thought wryly, surely a peer through the bush is worth two in the Lords.

Walking along those mock-medieval Gothic-tiled corridors, it crossed my mind I might well be walking in the footsteps of the great and the good from southern Africa's past. Maybe Sir Evelyn Baring had trod this very staircase. Remember him? Later Lord Howick? I imagined the satisfaction he must have felt knowing that his life's work bringing order to Africa had been crowned with having a dirt track in Piggs Peak named after him. Or what about the Earl of Beaconsfield, aka Benjamin Disraeli? His last government was brought down after the British army was soundly defeated by impis of that great warrior Cetcewayo during the Anglo-Zulu war of 1879. Then there was Lord Selborne, High Commissioner for South Africa who in 1907 issued the proclamation that Swaziland should become a Crown Protectorate, separate from South Africa. The Swazis, always able to catch an expression or a mood, had called him *'lisolibovu,'* old 'red eye'. Maybe he'd been at the colonial hooch.

Even His Majesty King Sobhuza himself had been here too, in 1923, soon after his coronation, to put the case for land ownership reform so that Swazis could own their own farms again. Fat chance! He had been met by the Duke of Devonshire, Secretary of State for the Colonies, who knew a thing or two about land ownership, possessing as he did enormous estates in England. There is a photo of Sobhuza sitting in London dressed in pin-striped trousers, a tail coat and polished top hat. It is a far cry from the traditional *incwala* regalia of ox tail pectoral and iridescent feathered headdress I had seen him wearing at Lobamba only a few short months before.

I arrived at the committee room in a daydream and opening the door, walked in. There they all were, the SARA steering committee. Who was it said: 'To get anything done a committee should consist of no more than three people, two of whom are permanently absent?' Well, I quickly learnt that this committee was no exception. It was the triumph of attendance over activity. Everyone had their say, and no one said anything. There was much made of the relevance of our ex-colonies, of encouraging the Commonwealth to develop its full potential. There was a lot of discussion about self-help, recognition of regional duty, responsibility to future generations, all three suggestions placed onerously back on the shoulders of the Swaziland government. The 'winds of change' had blown through Africa, but they had also blown through

Whitehall. The new message to Britain's ex-dependencies was, 'It's all terribly, terribly important, but we just don't have the money, so just push off!' I felt myself going backwards, not forwards. The committee turned out to be a master-class of inaction. If Henry Morton Stanley, the African explorer who did the 'presuming,' had had a steering committee he would still be shacked up in St Asaph workhouse where he was born and Livingstone would have disappeared without trace. The question was, now that I was in, how did I get out?

As the morning wore one, it became apparent that what the committee driving the SARA express wanted to do was to shunt the Swaziland Archaeological Research Association into a siding to await the slow train for Mbabane. In other words, they saw the steering committee as steering the problem right back where it came from – Africa. If all went well, and good things came out of the whole expedition, the committee would always be there in the wings, ready to take the credit. If it sunk into the mire, it was nothing to do with them. As they saw it, they had nurtured it to fruition and rightly handed it on to the proper authority in Swaziland. But what was 'the proper authority' in Swaziland? I left the meeting sadder and wiser. I wrote to Johnny who wrote back saying that there was a 'proper authority' we could use as the locomotive to regain the momentum. It was called the Swaziland National Trust Commission. That is where SARA would have a true home. I put the Lords behind me; who needs them, I thought? I'd had my moment of glory. Sobhuza himself hadn't really achieved much all those years ago at the 'Lords' and had gone back to Swaziland empty-handed and got on with his reign. I would do the same. I would assemble a team for the expedition, grab what grants I could under the auspices of the steering committee, and, as Ralph had said, just get on with it. Greatly heartened by Johnny's suggestion, I cast around to look for a team of experts. And as I pondered, our daughter was born, which rather changed the bidding somewhat. She turned out to be the most beautiful baby in the whole world, despite being born bald as a billiard ball with a head shaped like a hen's egg. But here she was. She could be the first new member of the expedition! I resigned my academic post, and with what meagre endowments I could raise we set out *en famille* for Africa again. I felt like Livingstone all those years before, leading his wife and children on some fervent whim into the African unknown.

* * * *

And so it was that the three of us returned to Swaziland. It was winter now and the face of the veld wore a desiccated mien. The mealies had long since been reaped and the parchment-coloured stubble stood short and rigid in the barren fields. The verdure and lushness of the summer had fled and left behind a sear and terracotta palette. In the highlands the dry, brown grass was burning. For days on end the rings of fire could be seen on the mountains. At night it was like a scene from Dante's Inferno as the flames leapt and died on the slopes of the Mdzimba, consuming the sapless shrubs in flares and scorching the earth. By day smoke drifted across the valleys in an endless pall. Where there had been green there was now a patchwork of burnt umber and charcoal black. A listless breeze gusted over the fields, whipping up the dry plough land into minor whirlwinds and blowing torn plastic bags against the barbed wire along the roadsides. The land had lost all its summer fecundity. The Swazis described this as the time when you ate the pickings of your teeth, when fresh produce was scarce and families survived on the maize they had stored in February and March. In the thatched homesteads of the middleveld the mealie cobs had long since been dried on the roof of the huts in the late summer heat and put away in mud and pole storage bins set high on stilts to stop the rats gnawing the winter supplies. The pumpkins and squashes sat on sapling racks, their chrome-yellow and manila colours blending with the ochre and dun of the ground.

But it felt good to be back. As we drove down the Malagwane Hill from Mbabane and saw the magnificence of the Ezulwini valley, I remembered our first view that previous December. It was no longer unfamiliar to me. It welcomed us back. Sheba's Breasts stood bare but imposing, beacons of optimism in this austere, pallid landscape. We went down to the bushveld. It lay in the withering grip of drought, the leafless trees standing stark against a pale sky. The impala moved about listlessly, feeding on the spent grass stems, flicking their tails and walking nervously across the exposed land, their cover gone. The rut was over and the males had fought their opponents for the domination of the herds. The females were already in calf and it would not be long before they dropped their young, but the dominant males were losing condition chasing away the young bachelors who were sniping around the edges of the group. The land looked dormant, wan and dying, as though its greenness would never return. But here we were, getting on with it. We had returned.

It was time to make a start. The team initially consisted of my wife Sue, myself, Ralph and now and again Johnny, and our two month old daughter. The first thing I had to do was to make my number with the Swaziland National Trust Commission, the fledgling body who had recently been constituted to oversee the development of National Parks, on the environment and wildlife side, and the National Museum on the tradition and heritage side. Right at this moment, they had yet to acquire either. At my first meeting with the Trust Commission, which at that time was located in a pokey office in Mbabane, I was introduced to its chief executive officer, James Mkhulunyelwa Matsebula. James had been and still was a famous man among the AmaSwazi. He was a great traditionalist and had been Sobhuza's private secretary, known as the 'king's eye.' He was from the powerful Matsebula clan who were allowed to be very close to the emaDlamini royal family on the basis that no offspring of a Matsebula wife could ever be made king himself. Even though his short curly hair was by now peppery at the edges, James was a handsome man, and, more especially, turned out to be one of the most gracious people I have ever met. He had an old world, old Africa charm about him, and he taught me more about Swazi lore and custom than anyone I came across. However, our first meeting was wreathed in confusion. I entered the room and he was seated behind his desk. I stood waiting to be asked to take a chair. He remained silent behind his desk waiting for me to sit down uninvited. This impasse continued for some time until he stood up, and coming over to my side greeted me in that particular Swazi way.

"*Sawubona, Dokotela.*" "I see you, Doctor!"

"Now, Dokotela, let me show you the way we do things here in Swaziland."

He then described the elaborate ritual of coming into a room. In order to be polite you should always try to be lower than the person you have come to see, in other words, shuffle in and sit down immediately, without being asked. I noticed this 'lower than you' performance many times after that. We would be chatting in James' office, both he and I seated, and his secretary would come through the door bent nearly double to try to be lower than both of us.

So here I was in the presence of the great J.S.M. Matsebula, Swaziland's eminent historian, and C.E.O of the Trust Commission. I felt I had to make a good impression. I sat down. Using the speech I had developed for the House of Lords committee, I immediately and eagerly galloped into my description of the forthcoming project. With a beatific smile he at once held up his hand to interrupt me.

"Oh, Dokotela, Stop! Stop!" he laughed, restraining me gently. "We haven't talked about your wife and your new daughter yet. And how is your father-in-law, Ingongoni? Is he well? He is still in Malkerns, is he not? How is he? And how is the farm? And his children are all well? And how did they manage in the recent cold spell? And how are you today?"

"I am well," I said, "but you know my father-in-law, Farmer Joe, Ingongoni?" I was rather surprised.

"Everyone knows Ingongoni, Dokotela. Swaziland is a small place and he has been here many years. And you are married to Ingongoni's daughter. You are indeed fortunate."

So we talked about the farm, about the family, about Swaziland today and yesterday. His eyes twinkled and he punctuated the discussion by nodding his head and pursing his lips in approval. He was truly pleased to see me, genuinely delighted to welcome me to Swaziland, like welcoming a new member of the family. He was interested in my views on how beautiful Swaziland was. We talked about everything except why I was here.

After a while, he paused. Then he asked if I would excuse him. He had to go to Lobamba. He had been summoned to a meeting with the king on a matter of some traditional etiquette.

"But we haven't talked about the project at all!" I said with some dismay.

"Oh," he smiled, "I am sure you know what you are doing and will tell us all in due course, when you are ready. *Hamba Kahle, Dokotela* – Go well!"

And with that he left. In Swazi society it was polite to be principally interested in the person, not in his proposition.

When I got to know James better, he used to call me *'lumfundzisi ematje,'* 'the man who can teach us about stones,' meaning of course the stone implements from prehistoric Swaziland. He would introduce me to his friends from the Swazi aristocracy, putting his arm around my shoulder and saying: "And this is a very, very clever man, Dokotela Price Williams, *'lumfundzisi ematje.'* He will tell us everything about our past, and the past before our past!"

It used to remind me uncomfortably of a story I once read about a geologist trying to explain the age of the gold reefs in the Witwatersrand to Paul Kruger, that ponderous leader of the back veld Boers in the Transvaal in the mid-19th century. When the earnest young man had finished, Oom Paul called to his wife and said, 'Mevrou, come and meet the man who was with

God when he created the world!' But James and I got on famously, and he chose his description of me with the greatest respect. He was a man of skilful traditional cunning, but not a jot of guile.

* * * *

The most important addition to the team and the fulcrum of the whole environmental concept of the expedition was going to be the geomorphologist, someone who could interpret changes in landform, the history of soils and erosional and depositional events. I had worked hard while I was in the U.K. to recruit someone who I thought would be the right man, or woman, for the job. I ended up choosing someone totally unknown to me. I had worked with two geomorphologists in Gaza. I contacted the first. He only 'did' the northern hemisphere. The other was stuck up a dune on the Coromandel Coast and wouldn't be back till a year next February, if at all. In any case, she specialised in coastal geomorphology, and conspicuously, Swaziland didn't have a coastline. Disaster! I called a geographer in Oxford and outlined the problem. To my amazement he said there were two or three people in his college who were at a loose end and he would get back to me. In a suspiciously short time, like two minutes, he phoned back and said that a certain A. Watson might be interested, and should he send him to London to see me? With some reservations I agreed and the following day we met in the Shakespeare's Head in Camden Town.

I recognised him as soon as he walked through the door, a strikingly good-looking, slim man in his early twenties with piercing pale-blue eyes and a shock of raven black hair. He was from Widnes, in the north of England, and spoke with a pronounced Lancashire accent. His father was a speed cop on the M6, and his mother, who was Austrian by birth, was a totally obsessed collector of 19th century Middle European model railways. With such a pedigree, how could we fail? He had just finished his doctorate at Oxford, so, my God, he was Dr Watson! Did that make me Sherlock Holmes?

The most noticeable feature of that afternoon in Camden was that Watty drank pints of lager like a tank engine on an incline. He was a career toper who if nothing else would fit in well with the All-Swaziland serial beer-drinking fraternity. By the evening, when I poured him onto a train back to Oxford, we were both legless. I can't remember much of the academic discussion about palaeoclimate in the sub-tropics, but I do remember that I hired him, several times. Watty turned out to be the best field-man I have ever worked with, the best I have ever known. In all the years we worked together, there is no question that with his forensic vision he was really the Holmes and I was the more pedestrian Watson. Watty understood landscape like no one I have ever met before or since. He just read it instinctively. And all the women on the expedition fell in love with Watty, with his handsome good looks and brilliant mind. He was in fact achingly shy most of the time, and hid his bashfulness behind a row of pint glasses. But every woman that ever met him thought they could redeem him and mould him to their wily ways. They all thought they could mother him, young and old alike. They all wanted to marry him, and one day, one of them actually did.

Months later I was turning over the events of that afternoon in the Shakespeare's Head, as much as I could remember of it, as I waited at the tiny airport at Matsapha, Swaziland's 'international air link.' The prop aircraft touched down and out came Dr. Watson. Our expedition was about to begin. I picked him up in one of our 'new' Land Rovers. In those early days we had somehow acquired a couple of old Series Two station wagons. They were known imaginatively as 414 and 417, after the number plates, SD 414 and 417. They had the same missing components as the farm vehicles, the non-essential bits like steering, brakes, and such like, but we loved them dearly and they stayed with us throughout our wanderings.

Our first foray some days later took us down to the Mkhondvo valley in the centre of the country. It was an especially beautiful valley, with a meandering river surrounded by majestic sycamore figs set between granite mountains wooded with kiaat trees and giant euphorbias standing up like monstrous cacti. Watty had studied the aerial photos of the Mkhondvo and had noticed some exposed patches hidden in the bush. We drove along the dirt road looking closely through the trees until finally, on a shallow slope, we found what he had seen on the pictures. Parking the Land Rover we walked off into the undergrowth. The trees opened out and there in front of us was the most extraordinary sight, a treeless gully broadening upstream into a huge gallery of bare, vertical flutes of earth, like a series of organ-pipes, all bright white in the sun. The light reflecting from these was so strong we had to shade our eyes. We learned later that this whole feature was known as a 'donga.' The bottom of the donga was a sinuous sandy bed, and as we walked up the middle the organ-pipes crowded round offshoots of the main arm in pleated cliffs many metres high. Watty looked intently at the walls of the donga, picking at the organ-pipes with his geological hammer and playing with the hard earth. Half an hour went by.

"Bingo!" He suddenly exclaimed. "See this crap here? This gully is cut through some kind of sediment that's been emplaced here by erosion from that back slope." He pointed to a group of granite koppies (hills) higher up the side of the valley. "You see, it's currently eroding, so it's not filling in at the moment. That means there is a cycle of deposition and erosion, and if we are lucky, that will be climatically controlled. I wonder how old it is?"

Just like that! He had almost cracked the climatic code on day one though it would take us up to ten years to find out everything we needed to know about these dongas. There was something else. Scattered all over the floor of this donga, and eroding out of the walls, were masses of tiny man-made flakes, and among them beautifully-flaked stone arrow points and scrapers belonging to the Middle Stone Age, from tens of thousands of years ago. I showed Watty, and sitting down on top of the organ-pipes I described the sequence of the Stone Age and the newest dates that had been ascribed to it. We couldn't as yet guarantee the exact relationship between the organ-pipes and the flakes, but the chances are that in some way their dates were related. Over the next few weeks Watty found more and more dongas, checking with the aerial photos and marking the bare patches on a map. Driving '414,' he covered the whole of the Mkhondvo Valley, and valleys nearby.

Watty had a way of driving a Land Rover that was unique. He recognised only two speeds, stop and flat out. The tales of Watty and his hair-raising drives across Swaziland, roads or no roads, could fill a small manual of dos and don'ts for prospective 4X4 owners, mainly don'ts. But he had his opponents and his *bête noir* turned out to be Swaziland United Bakeries, SUB. He had a growing obsession about this company bordering on the manic. Watty would describe how he had been driving alone down a long pediment slope of a river valley on some remote dirt road miles from anywhere, hours from the nearest civilization, when there, at the bottom of the slope, as though on a starting grid, would be an SUB truck. According to Watty, it had obviously been there all day waiting for him to show up. The moment he came round the bend at the top, an imaginary chequered flag would wave and the SUB truck would accelerate towards him at speed, 0 to 60 in so many seconds. It came flying at him in a cloud of dust up the crown of the road. Neither would give way. Watty would hold his own, but the diminutive Swazi behind the wheel of the SUB truck was not to be intimidated. He was coming straight for him. At the very last moment Watty would give in and lurch off the road into the ditch, cursing royally. He never won in all the years they jousted together, SUB and Watty. No one else ever suffered this affliction, but if you ever wanted to get Watty upset, you just had to mention SUB.

Watty located many more dongas, and one by one I went to see them with him. They all had the same features, the organ-pipes, the sandy bottom, the Middle Stone Age flakes and the shaped projectile points. They were all in distant parts of the country, in difficult places to get to and there were hundreds of them. Some were compact, no more than a few hundred yards long. Some were shallow, with the organ-pipes planed down by heavy erosion. But some were huge, up to half a mile or more long, and deep, anything up to fifty feet or more. The earth of the pipes was hard and unyielding. But tiny stones eroding out of the gullies lay on the lower surfaces like ball-bearings which made walking treacherous. It was like treading on a roller skate, and on many occasions Watty and I would return home with bleeding legs and elbows. But we were making progress.

CHAPTER EIGHT

UNPRONOUNCEABLE SCHOLARSHIP

Kunhlobonhlobo.
> siSwati. Translates as 'of various different kinds.'

We were staying at the old farm in Malkerns at the time. One night, after a long day down the dongas, Toko, our maid of all work, came into the dining room where we were finishing supper and said there was a problem. There was a Swazi woman by the back door who was pregnant and was about to give birth, but there were complications. Watty and I were doctors, she implied, couldn't we do something? Much fine wine having been taken, we were already in a state of semi-inebriated gallantry. In those days we drank wine out of a box with a five litre plastic bag inside which caved in as the wine was used up. Generically, we called it *'vinho collapsico.'* We called the white wine 'the strong dog,' for obvious reasons. The red tended to lend extra vigour to our nightly discussions and was 'the red infuriator.'

"A flute of the strong dog?" we would enquire of each other. "Or would you prefer a glass of the infuriator?"

That night, after a heroic intake of the 'infuriator,' everyone was aggressively hearty, if not downright belligerent. But the woman was desperate, moaning plaintively at the back door. Watty and I jumped into 414, bundling the poor unfortunate patient into the back, and drove off 'Watty style' into the night. I half expected to see an SUB lorry around every bend, but the drivers must all have been asleep at home by now, getting ready for the next day's fixture. As we rocked and rolled along the dirt road, the woman's groaning became louder and more desperate. I couldn't tell if it was because of the imminent birth, or that she didn't appreciate Watty's driving technique.

"Who is she anyway?" asked Watty.

"I don't know," I said, "I've no idea. I've never seen her in my life before. Neither has Toko."

We drove on through the night. By her rasping wails the woman was clearly approaching the crisis. For all I knew she was about to be delivered on the floor of the Land Rover. It was touch and go back there on the new 'rear pimple matting, rubber,' Land Rover spare part no. 45/6/9995T. I conveniently knew the part number because we had just put it in, so if she did pop, that would be the 'place of birth.'

After what seemed a geological epoch, we reached the outskirts of Manzini, 'the Hub,' and skidded round into the forecourt of the Raleigh Fitkin Memorial Hospital run by the Church of the Nazarene, a worthy organisation from Kansas City, Missouri. We opened the door and bodily lifted the woman into the reception area.

"Name?" asked the stiff-collared sister behind the desk. She was all starch and floor polish, this one.

"I have no idea," I replied.

"Come, come!" she scolded. "Even in your intoxicated state you can remember your own name, surely. It's the one your mother gave you!"

"Listen," I wheezed, jabbing my finger at her, "it's not me that's the problem. It's that woman over there. Can't you see, she's about to have a baby."

I pointed over my shoulder. The woman was prostrate on the bench with Watty bending over her.

"So who's he? The father?"

The infuriator was beginning to kick in well by now. "Neither he nor I have any idea who this woman is, and the child, if it ever gets here, had nothing to do with us either. Do I make myself clear?!"

Sister was not to be intimidated by a couple of donga wallahs like us, tanked up on the *vinho collapsico*. She gave me a withering look as if to say she didn't believe a word of it.

"So who is going to pay her fees?" she continued levelly.

Ah, snag! I was 'without funds,' I knew that. Watty rummaged about in his shorts, never the most salubrious at the best of times, what with the organ-pipes and all, and dug out a few malodorous crumpled notes.

"How much does she need?" he asked expansively. The woman next to him was by now comatose.

We found just about enough for her confinement, and handing it over the counter, walked to the door to leave.

"And what is the baby's name to be?" enquired the sister, to no one in particular.

"Christ, I don't know! Call it Solly!"

As we drove away, Watty asked, "Why Solly?"

"You know, after Solihull, Warwickshire where they make Land Rovers."

"Suppose it's a girl?"

"Then she'll just have to be called Elizabeth," I said, "after Port Elizabeth. Isn't that where the South African Land Rovers are assembled?"

It turns out that the baby was born that night, and mother and child did well, as far as we know, but whether she called it Solly or P.E. we never knew. I do know that a year or so later, when our Toko herself gave birth, she called her first-born son after Watty. I told you all the women fell for Watty.

* * * *

In the weeks that followed we went through innumerable dongas all over Swaziland, way north along the Poponyane river beyond Piggs Peak, south passed Sitobela on the Mzimnene river and on the lower Komati. Everywhere the scenery was striking, even in the dry depths of winter. Though the land lay empty and the cattle had become lean for want of richer fodder, there was a calmness about those warm, rainless days. It was exhilarating to travel from one new site to another, not sure what we would find, learning our trade as we went along. In the Mashila donga, we walked into it from the river through a narrow high-sided channel. The vertical walls rose four metres or more above us as we squeezed through into the main gully which opened out into a huge theatre-like auditorium with pinnacles and organ-pipes all around. Tree wisterias were in bloom all along the edges, their pendulous delicate violet pea-like flowers contrasting strongly with the white walls of the gully. Along the donga floor prehistoric implements lay scattered on every surface. In one case there were even pieces of iron oxide with rubbing stones, and I was reminded of Lion Cavern at Ngwenya and the mining for pigment all those thousands of years ago. It seems that when this sediment was accumulating on these surfaces, the hunter-gatherers of the Middle Stone Age must have sat here, near the river, and made their tools. In some cases we could match the flakes and fit them back together, evidence of a single tool-making moment in the past.

Watty loved to find the most obscure siSwati names for the sites, on the basis that when it came to publishing a scientific paper, once these names were 'in the literature' everyone had to adhere to them, causing great hilarity at international conferences. There are a number of exotic sounds in siSwati which had been borrowed from the San Bushmen during what must have been a thousand years of interface between the Nguni cattle keepers and the San, last of the Stone Age hunter-gatherers. The Nguni speakers adopted some of the clicks that you can still hear in Bushmen languages, like !Kung, today. So some of the Swazi place names make very interesting hearing, and even more interesting spelling. The more phonetically difficult they were, the more Watty liked them. So we had scientific sites he named Ethengethengeni, Ntabamhloshana and the best of all, N!kon!wane, which sounded like a game of skittles by the time you got the pronunciation right. He would roll these around his tongue as he sampled the soils, savouring the day when someone else would have to repeat them.

"Well," he reasoned, "it's only the same as these lunatic soil scientists, you know, the ones that get to call clay soils by these crazy names. Everything is some kind of '-onite' in soil science, so you get names like Uptonite, Intonite and the worst, Cummingtonite. You feel such an idiot having to repeat them. Well, let them try Kunhlobonhlobo next time!"

* * * *

On field days we would usually stop for lunch at a river ford in the bush, now no more than a winter trickle flowing over the sand between granite boulders. Wild date palms grew along the edge of the stream. As we sat on the front wings of the Land Rover, the slow pace of rural Swaziland went on around us. The local people would greet us with the simple and lovely greeting:

'*Sanibonani bonkhosi,*', 'I see you, friends,' each one lifting both hands in front of them almost in a blessing.

My reply would be

'*Yebo, ngibona wena.*' 'Yes, I see you too.'

Everyone we met accepted our presence there as we walked up and down the shallow valleys of the middleveld. The immovable cattle looked on with a cow-like lack of interest.

This was the time of the year when the women collected the tall *hyperenia* thatch grass that grew along the edges of the bare mealie fields, and they would go out to cut it with sickles balanced by the edge of the blade on their heads. It certainly tested the deportment. They preferred to carry things on their heads rather than in their hands, so you would often see a Swazi woman in traditional red wrap and full length black skirt walking through the bush with a saucepan of water or plastic bowl of laundry on her head, or a bottle of beer or small tin of meat, or even a loaf of white bread bought from the local store four miles away. SUB must be around somewhere, Watty reasoned.

Our lunch usually consisted of bees flies. That's what it said on the tin, 'bees vleis,' which turns out to be the Afrikaans for corned beef imported from the Republic. Also we carried with us a sack-full of enormous avocado pears each the size of a grapefruit which we could buy for next to nothing in Mhlanyana market near Malkerns. These we ate with a desert spoon with plain salad oil and vinegar. As we finished, the women who had been collecting the grass would wend their way back past us, huge bundles of thatch sitting long-ways on their heads. This was rural Africa at its most traditional. It was all a long, long way from the House of Lords, for which much thanks.

Now and again a beaten-up bus would come crabbing down these remote dirt roads belching black smoke and sending up clouds of dust. This was another bane of Watty's driving career. He maintained that it was company policy for the drivers of these buses, if they saw Watty driving his Land Rover in front of them, to overtake him and then make an emergency stop, with the lame excuse that this was the next bus stop. The bus companies were quaintly named in those days, though where they had got the titles from is hard to fathom. There was one called the Tit for Tat Bus Company from Siteki, on the Lubombo, frequently to be seen on remote bushveld roads. There was the Ding Dong Bus Company from Manzini, along with their competitors, the Green Sea Transport Company. Perhaps the most obscure was the Eat and Cry Bus Company. But the one that got up Watty's nose more than any other was the Muhle One Way Bus Company. *Muhle* means beautiful in siSwati, but why 'One Way' we shall never know. They might be beautiful to some people, but to Watty they represented automated hell. Many an evening he would come back from a long day in the farthest recesses of Swaziland, fresh from an ugly scene with a 'Beautiful One Way' bus that had narrowly missed killing him near Dvokolwako or some other improbable place.

In the bushveld the wild pear trees were in full flower. They are not related to the pear family at all, but they have showy sprays of the most attractive, creamy-white blossoms. Alongside them were the lucky bean trees that also bloomed in the winter, sending out the most strikingly brilliant spiky scarlet flowers that emerged before the leaves. To see one against a deep blue sky was almost a psychedelic experience. It is called the lucky bean tree because it drops vividly coloured black and red beans after the flowers have faded. The Swazi girls used them for making necklaces. You could buy them in the hotel shops in Mbabane, leading to the intriguing fact that when white people came first to Africa they traded in beads for what they wanted - copper, ivory and gold. Now the Africans are selling beads back to the white tourists.

The more Watty saw of the dongas, the more impressed he was with them. As he mapped out the extent of these sediments he discovered that they covered about a third of Swaziland. Some dongas even displayed an earlier, bright red sediment beneath the white. These ones we called 'two-tone' dongas and every donga had stone tools in it. The two-tone dongas often had hand-axes in them, from an earlier phase of the Stone Age. These gullies were becoming extremely important to us. They had great dimensions in time and space. Agriculturalists in other parts of the continent had written dire warnings that gully erosion was eating away at the very fabric of Africa and its rural economy.

"Don't tell anyone," Watty confided, "but we are the only scientists in the whole of Africa that actually love soil erosion! Isn't that awful?"

The more he looked at it, the more he suggested that what this sediment consisted of was colluvium, a slope sediment sitting on these low angled pediments leading down to the main river courses. In fact, it was a pediment sediment. That and other *bons mots* helped to pass the time. 'What's red and white and sounds like a bell?' 'A two-tone donga!'

It was time to take our findings to Johnny at his estancia in the 'Place of Winds.' Johnny, Stella and the celebrated farting dogs had met Watty when he had first arrived but since then we had spent our time 'in the *bundu*.' We drove up one evening from Malkerns in time for dinner. We began our explanation over a few libations of Old Buck gin and tonic. Being good Aberdonians, Johnny and Stella drank whiskey, in particular a peaty concoction smelling of burnt rubber and called Old Toenails or some such name. I began by describing the location of the dongas and the heaps of stone artefacts that seemed to be washing out of the colluvium. I showed him some. Watty then laid out the processes by which all this might have taken place. Johnny was impressed, but interrogated us very closely. He seemed to understand all our weeks of work in a trice, and had already formulated the questions we ourselves had asked. The dinner was excellent, accompanied by fine wines - these wines really were - and Watty was helped to more fillet steak, more vegetables, more desert, until I realised that Stella was taking more than usual care over Watty's welfare. "No! No!" objected Stella later, "it's just a motherly interest, you know. The poor wee chap needs building up, all those dongas and such like!" Bugger! Inadvertently Watty had done it again. It was those ice-blue eyes and raven hair.

Next day we were in the *bundu* again. I was anxious to show Watty the terrace with the hand-axes on it. Down on the Ngwempisi we walked the same fields I had walked with Johnny, and sure enough the hand-axes were there again. Watty paced the ground, walking down to the river, then up the slopes again. The more he looked the more he frowned and rubbed his chin. He picked at the exposed sections and felt the silt between his fingers. We ranged far that day in the Ngwempisi valley until quite suddenly Watty turned to me and, with a small two-step, whacked his fist into his other palm

"Cracked it!" he shouted.

"What is it?" I asked. "A river terrace?"

"This isn't just one river terrace, Sunshine! It's a series of terraces going back through time."

We walked up and down the valley and he pointed out the various features the river had left behind after what turned out to be its long and chequered career. The first was a sandy terrace, low down near the water. This was recent, Watty said. Above that was a grey silt terrace, probably from the last few thousand years. And so on, back up the slope. Watty outlined the whole history of the river, back to the hand-axes and beyond. It was truly riveting. We wrote all this up some years later for a scientific journal. Watty was especially pleased with the name he chose for one of the terraces which sat uniformly about thirty metres above all the rivers in Swaziland. He called it the 'Shongololo Terrace'. *Shongololo* was the siSwati word for the huge centipedes we used to see worming their way through the bush.

"See, they'll all have to use it now!" he triumphed. "See what I mean? Centipede? A hundred feet above the river!"

Watty might be cracking up, but slowly we were cracking the code of the geomorphology of Swaziland. We drove back to Malkerns. This would be music to Johnny's ears.

* * * *

Academically, we were moving at great speed. Physically, with Watty driving, we were moving at great speed as well. It was a time of speed traps, and we were about to encounter one. Believe it or not, despite Swaziland being a traditional kingdom, a real African backwater with dirt roads, some bright spark in the Swaziland police had suggested they should buy some speed traps, and from humble beginnings speed trapping had turned into a national sport. The supposed reason for these instruments was to reduce the appalling carnage that took place on Swaziland's roads every month of the year. It is true, the level of accidents was seriously dreadful, and the weekly wreckage along the main roads made stock-car racing look like the State Opening of Parliament. But it was noticeable that the police were always out with their speed traps just before the end of every month. Watty always maintained that they had a hidden agenda; he felt sure they were trapping the unsuspecting motorist and issuing on the spot cash fines to enhance their own coffers just before pay-day. True, that was when most of the accidents happened, not only because that was when everyone else got paid and people could afford petrol for their *isikorokoro*, the local name for an old banger, but that was also the time for the monthly celebration. It was party time. Whatever the reason, out would come the double wires across the road, especially on a sharp down-hill stretch of road just on the edge of town, when by looking at the scenery you would have thought you were still in the *bundu*.

The speed limit in towns, such as there were, was sixty kilometres an hour, and after a while everyone knew where the speed traps would be and helped each other to avoid getting caught. Once, we were coming down a steep incline into Manzini and oncoming drivers had warned us that the trap was there. Lights had been flashed and arms waved through the windscreens and sure enough we saw the lines on the road and eased our way over them. Admittedly the speedometer on 414 worked in an erratic manner and frequently it would show that we were travelling at one hundred and twenty five kilometres an hour when we were barely moving, but we had got used to that. We were by now fairly good at judging what sixty kilometres looked like, so over we went. Just down the road the smartly-dressed officer in charge boogied across the road to stop us. He had caught us fair and square this time and we would have to pay. He forced Watty, who was driving, to get out and to come over to the box, where the offending speed had been recorded. It read sixty one kilometres an hour! The constable gave Watty the grand remonstrance about reckless driving and demanded some exorbitant fine. Watty turned puce! He started haranguing the officer, describing elements of the human reproductive system, and even mentioning specific organs! With the aid of several references to the lower digestive tract, he pointed out that their machinery was totally incapable of being that accurate, especially on a dirt road. It may not have been the most edifying soliloquy, but what it lacked in eloquence it made up for in raw energy. I was under the seat with mortification by the time he had finished, expecting to have the Land Rover impounded and ourselves locked away in the slammer in Matsapha. But no, Watty had made his case. The officer was so taken aback by Watty's forceful logic that with a huge toothy grin he wished us well and, apologising for breaking our journey, waved us on our fine-less way.

* * * *

The next week we spent time on the lower reaches of the Mzimnene near St Philip's Mission, one of the many remote Christian missions that are spread throughout southern Africa. I am sure they did good works and helped the local community, such as there might be in these under-populated parts, but it was always a surprise, after driving for hour after hour along deserted bush tracks to come across a fully-fledged church in the wilderness. One of them, with rather startling incongruity, had been named St. Martin's in the Veld. The corrugated iron facade was more likely to look out on actual lions than the Landseer lions in Trafalgar Square, we thought. There was even a mission on the Great Usutu upstream of Malkerns, the one-time incumbent of which had been the Rev. Molesworth, known ubiquitously as 'Holy Moley.'

On the Mzimnene River at St Philips there were large expanses of very early gravel terraces spread high above the present course of the river. These were littered with stone artefacts, some of which came from a very remote period indeed. They were river pebbles that had been modified by removing only a few flakes at one end, and the resulting scars were deeply stained, which offered further proof of their great antiquity. When I checked what they were, they seemed to be identical to material that had been found at Olduvai Gorge in East Africa. If so, then this was going to be phenomenally important. Olduvai lies below the western slopes of the Ngorongoro Crater in Tanzania, made doubly famous you will recall by the ever-lovely Michaela Dennis, she of the open bush shirts. Olduvai is right on the edge of the Serengeti Plains. A huge gully has recently cut through this area and exposed sediments as old as two million years ago, at the very dawn of Mankind. As a result, Olduvai Gorge has become one of the most famous archaeological sites in the world. It was in the base of those sediments at Olduvai that the celebrated couple Louis and Mary Leakey had discovered stone choppers made from river pebbles that were reliably dated at 1.8 million years ago, some of the oldest examples of tools that had been found anywhere on Earth. Mary Leakey had spent much of her life analysing these artefacts. Her work was exemplary, accepted world-wide as describing one of the true beginnings of human technology. At this famous site of Olduvai she had itemised five hundred artefacts here, six hundred or so there, and named the whole assemblage the Oldowan Industry, after the Gorge. They were globally iconic, both the tools and the Leakeys.

To my eye the artefacts on the Mzimnene terrace belonged to this same period and this same industry, and indeed that eventually is what they were confirmed to be. They were Oldowan. The reason they had survived for so long is that they had been made from the most resistant rocks we had ever come across, the Mozaan quartzites, which rendered them impervious to weathering or decay. So here they lay, out in the open. They emanated a sort of technological magnetism. Holding one of these crude pebbles in your hand was like touching the dawn of humanity, vaulting back to the very beginning of human existence. Yet there seemed to be so many of them. Watty and I criss-crossed the terrace for some hours. I drew and classified some of the choppers while Watty measured their height above the river, checked the geology and generally described the whole feature. Around about lunch time, with bees flies and avocado to look forward to, I saw him loping back to the Land Rover where I was writing up my notes.

"Have you any idea how many artefacts there are on this terrace?" he asked. "Give a guess, go on!"

I hadn't a clue. I had spent my morning in only one area really.

"Well, wait for it! At a conservative estimate, and I stress conservative, I would say about one point two million metric tonnes per square kilometre. One point two million metric tonnes! That would make tens of millions of artefacts. How many do they have at Olduvai Gorge, then?"

I looked it up. "About three thousand!" I croaked. "In total!"

"Jesus! Is that so? What the hell are we going to do?"

I mentioned our Oldowan problem some time later to an archaeological colleague who I thought might be able to help. I described the site and told him the scale of the challenge. We had a site many, many times bigger than anyone had ever seen before. What were we to do?

"If I were you," he replied, "and knowing the Leakeys as I do, I would go away and think about something totally different. It's far too big to worry about."

But Watty and I did publish the site, in brief anyway, as an example of the way in which landscapes had changed in the Swaziland Lowveld over the last couple of million years. We thought it was a valuable contribution to the sum knowledge of planetary science. It came out in a specialist German *Zeitschrift*. We had wanted to call the paper 'A Load of Old Cobbles!' but the humourless editor, from München-Gladbach, or wherever, had given it a common sense title, something about 'Preliminary Investigations into Early Pleistocene interfluvial gravel deposits in South East Africa.'

We went back to see Johnny with our discoveries, to the Place of Winds. He was elated. It meant that we had an archaeological record of human presence in Swaziland equal to anything they had found anywhere else in Africa. We had so much material and so much information it was difficult to know which way to turn. Swaziland was exceeding my expectations by a long, long way.

"What are you going to do now?" asked Johnny.

"Get help!" we chorused.

CHAPTER NINE

TALKING TO THE TREES

> *I talk to the trees,*
> *But they don't listen to me.*
> *I talk to the stars*
> *But they never hear*
>
> Alan J. Lerner

Help came, at least in part, in the form of a lanky youth named Colin, of whom more in a moment. News of our success in the dongas of Swaziland had reached as far as Oxford University, and with this in mind we approached a senior don to come and verify our discoveries. Andrew had worked in the bush in East Africa, amongst other places, as well as in the Thar desert in India. As he said:

"If I can survive in a piss-pot place like that, I can certainly make do in Swaziland."

He was a man of immense academic drive, much given, especially after a glass or two of the 'infuriator,' to firing off an endless string of suggestions. Why hadn't we done this? Had we thought of that? Had we read this paper, or talked to that geologist? Why weren't we sampling every so many metres instead of the way we were doing it? Why not try a different approach here, or a new technique there? It was all very helpful, and kept us on our mettle.

Watty and I took him to one huge exposure of colluvium in which we were particularly interested. It had a myriad lines across it, different coloured layers and different hardness of sediment. We felt it may throw light on some of the time-related questions we needed answering.

"Surely we must be able to interpret this," I queried, "or I'll take up accounting!"

Two hours later if anything it became more incomprehensible than ever. Andrew turned to Watty, as I poked around in the gully and said.

"There he is, archaeologist extraordinaire, accountancy staring him in the face."

It was not always easy, especially in those early days.

It was Andrew who suggested that he had a potential doctoral student who might be a useful addition to the team, a man who could consolidate our knowledge of the dongas. We were sitting around the fire at Johnny's one evening in 'The Place of Winds' during our second winter season. The dogs were asleep in the warmth. Never has the phrase 'let sleeping dogs lie' been more appreciated. We had ranged over the whole project to date and Andrew talked about grants that might fund a student for a three year period.

Out of this conversation came Colin. I first met him in Oxford, in the King's Arms at the end of Broad Street. He was a gangling, slender lad with 'Easy Rider' spectacles. At twenty one, already slightly thin on top, he exuded a likeable if nervous enthusiasm. He had just been awarded a starred first class honours degree in This and That. In other words, he was extraordinarily bright, though to me he came across as slightly gauche.

"Perhaps, Colin," chivvied Andrew, "you would like to begin by telling the director what you have been doing."

It had become something of a joke that I was the director of the Swaziland Archaeological Research Association, so Andrew used that title for me in a slightly mocking but not unkind way. Colin took it to heart.

"Well, director," he began. He outlined the degree courses that had led him to his heady result, but then, slightly obliquely, he began to talk about all his class mates who had just finished their degrees at the same time. Julian had just landed a super job in the City, earning absolutely scads of money. Sebastian had been commissioned into the Guards at some huge salary. His own closest friend Dominic had just gone into the Bank and within a year or two promised to be making millions. How would this genius with his massively ambitious friends get on down a donga in Phuzamoya, I wondered?

But the next season, out he came. He had read copious amounts about the country and its background and was a walking encyclopaedia when it came to bookwork. Still, with a starred first I suppose he ought to be good. We had extended our fleet by buying a new Toyota Hilux 4x4 truck, a bakkie. Brand new, it had all sorts of features that Watty and I could only dream about - comfy seats, cigar lighter, new spare wheel, brakes, heater, demister, everything really. That was the good news. The bad news was that we had bought it for Colin, which somewhat dampened our enthusiasm. We could hardly begrudge it to him, as it was mostly his money. By this stage we had managed to take over a comfortable house on the Swaziland government

Agricultural Research Station near Malkerns. From there Watty introduced Colin to Swaziland, showing him the dongas and the problems we were encountering after which Colin began his field work and he took to it like a duck to asphalt. After the third week he hadn't moved.

"Ah, yes, director," he prevaricated, "I've just got to get the plan of action right. Don't want to waste valuable time going in the wrong direction, do we?"

Another week passed.

"Colin, any chance of going in any direction, right or wrong, do you think?"

"Certainly, director! Actually, just finished the plan of action. Ready to roll, really."

And sure enough, the next day Colin drove off. I breathed a sigh of relief. An hour later he was back.

"Just been to Malkerns store to get more pencils. Need lots of pencils on a job like this!" he said, in all seriousness.

I lost my composure.

"Colin, will you get out there and start measuring dongas before I lose my composure. You're supposed to be doing a PhD, for God's sake!"

"Absolutely, director! DPhil actually, but I know what you mean. Start right away. Sure. Tomorrow, in fact. OK?"

And with that he did indeed begin. He would leave every morning and come back the same evening. Where he went to I had no idea, but the Toyota always looked immaculate. I would not have put it past him just to drive round the corner and sit there reading romantic novels.

There was something slightly inept, slightly disengaged about Colin. He was an attractive and charming man but somehow he always managed, with the very best will in the world, to say the inappropriate thing, or ask the eccentric question. That winter a toad had come into the house and was living, as far as we could tell, behind the bookcase. Every evening while we sat discussing the day's affairs this amphibian would let out a periodic belch. We were into the gins and tonics one night and asked Colin to bring in another case of tonic from the back of Watty's Land Rover. He banged it rather carelessly down beside the bookcase. The following evening our personal toad was silent. The silence continued the next night, but was replaced by an unpleasant smell that grew stronger each day, until, lifting the case of tonic, we found that Colin had flattened the poor creature. Ever after that, whenever Colin was pouring drinks before dinner, he would ask:

"Another gin and toad for you, director?"

* * * *

One weekend we took Colin with us to Malkerns Country Club (members only) to get him more into the swing of things. Colin fell to chatting to some folk at the end of the bar. I got talking to an old pioneering farmer about his own experiences. He was drinking the Transvaal elixir – brandy and coke. He had spent all his life in the isolated backveld of South Africa and spoke with a laboured English accent. When he found I was from the U.K., he had to tell me the sorry story about the only time he had gone overseas.

"I 'ad 'eard so much about U.K.," he began, speaking in slow, ponderous tones. "Everyone I knew said U.K. was the place. Man, they described it like a kind of paradise, you know? Well, I often thought to myself, if that's true, then what the 'ell are you all doing 'ere! But I kept my mouth shut. Anyway, last year I was persuaded by my wife to go and 'ave a look, eh? So we went to London, you know? 'Er and me. Man, it's a long way, you know. Any 'ow, we get there and we see everything - Buckingham Palace, Tower of London, 'Ouses of Parliament - everything. Anyway, there was one thing we 'adn't done at the end of that trip. I 'ad been told so many times that the place to go was Soho, you know, for the girls and that. They told me we don't have anything like that in southern Africa. So I told my wife to bugger off back to the 'otel and I went to Soho. 'Ell Man, those strip clubs, eh? They everywhere, you know? So I choose one and I go in. It's all dark inside. But it's not really a strip club, it's what they called a peep show. You put your money into a slot and a 'ole opens up and you see the girls on the stage, like dancing and that. So I put this pile of money in the slot and there on this stage was this black girl taking her bloody brassiere off and wiggling 'er tits at me, a whole five minutes worth. Iusus, Man, I got 'undreds of girls 'oeing my fields every day of the week dressed like that. Why would I want to go to the other side of the world and pay to see it? What a country, eh? U.K.? Who needs it? Cheers Man!"

I looked round to find Colin. He was trying in his own earnest way and his rounded Oxford vocabulary to explain what he was doing in Swaziland to a red-faced Afrikaner rep. for some South African chemical company who, by his glazed appearance, had probably been propping up the bar since the last time we were at the club some months ago. It was an uphill struggle for Colin since it was obvious that Pleistocene geomorphology was not part of his listener's stock in trade. Colin stood head and shoulders above his audience, but rather emaciated, like a coat hanger. As he reached the crescendo of his account, the Afrikaner, clearly bored stiff by now, took a thoughtful pull at his brandy and coke and turning to Colin and looking him up and down, poked him in the chest and said:

"Christ, Man! You like an African dog, eh? All ribs and prick!"

∗ ∗ ∗ ∗

By now we had started to map all the gullies in Swaziland and were beginning to wonder how far the colluvial story extended beyond the boundaries of the kingdom. We had picked up a clue. Sue and I, on our previous trip to the U.K., had seen a new film about the Anglo-Zulu War of 1879. I was becoming totally absorbed with southern Africa, ancient and modern, and this film was a re-enactment of the opening battle of that war, the same morning as the later battle at Rorke's Drift. Remember 'Zulu'? This was the prequel, 'Zulu Dawn!' The battle was called Isandlwana and turned out to be the greatest Zulu victory, and the greatest British military defeat.

With a cast of thousands, and Chief Mangosuthu Buthelezi playing his distant forebear, the Zulu King Cetshwayo, John Mills was Sir Henry Bartle Frere, governor of the Cape. Peter O'Toole played Lord Chelmsford, the general in charge who was responsible for all the cock-ups that led to the disaster. Six companies of the 24th Regiment, mainly drawn from South Wales, died in the ensuing battle, with hosts of others, among them Colonel Anthony William Durnford, Royal Engineers, commanding a group of Natal Native Contingent. Durnford was an interesting character. He was fascinated by Africa and was much loved by the Africans under his command. 'They are fine men,' he wrote in a letter home, 'all naked and that sort of thing, but thoroughly good fellows!' In the film Durnford was played by Burt Lancaster with an Irish-American accent. His portrayal of Durnford was a bit too swashbuckling for the real Col. Durnford, who was a somewhat upright Victorian colonial officer.

On the day of the battle, 22 January 1879, a Zulu army of twenty thousand made up of Cetshwayo's warriors came swarming over a ridge above the rock of Isandlwana and had almost surrounded the camp when Durnford rode with his Basutos and tried to turn the tide on the left flank by blazing away at the massed ranks of Zulus. But it was too late. After a heroic last stand, he was overwhelmed and stabbed to death by a forest of assegais. In the film the director panned in to show Burt, mortally wounded, falling backwards into a gully. Right there on the screen where he was supposedly breathing his last was a two-tone donga. On the film you could even see the organ-pipes.

"Look! See! It's a donga. He's dying in a donga! It's even got red and white colluvium!"

I was ecstatic. If the film location was correct, there were many more dongas in Zululand. Back in Swaziland I discussed them with Watty and suggested we went to have a look. We had somehow acquired a caravan that was currently parked in the garden under an arboretum of Cape honeysuckle so I suggested that we hitched it to one of the Land Rovers and set off on a field trip to Zululand. However, hitching up the caravan turned out to be more difficult than it looked. The actual ball and socket attachment was easy enough, but when it came to wiring the electrics, it was another story. We had six wires coming from the vehicle but only five coming from the caravan. We pressed the brake and the reversing lights came on. We turned right and the left hand indicator flashed. We turned left and the stop-lights came on.

In the end, we gave up and hitched the caravan up to Colin's pre-wired Toyota. He would have to come with us. We toiled up through the mountains and crossed into the Republic. As we left the border post behind on the South African side Colin became puzzled.

"I say, director, I thought South Africa was an apartheid government. What are they thanking the Africans for? On that notice back there, it said in big letters, 'Thank You, Darkie.' What's that all about, director?"

"Colin, it wasn't 'Thank You, Darkie.' We've just left the border post and they were expressing how grateful they are for our visit, in English and in Afrikaans. It said 'Thank you; Dankie.'"

The upland panorama of the AmaZulu Nation, the People of Heaven, is of singular magnificence, broad sweeps of country alternating with flat-topped mesas and rocky knolls. This is the land of Blood River, Majuba, Spionkop,

Hlobane and other battlefields which pepper the accounts of emergent South Africa over the past couple of hundred years, battles between the Boers and the Zulus, the British and the Boers and the Zulu and the British. Through it all the People of Heaven suffered and survived, their round mud and thatch homesteads still scattered in kraals on their wide plains and over their native hills. Even now the Zulus, only once defeated, by the British, still maintained a fearsome independence, their warrior image undiminished. In the deep recesses of the Tugela Valley, they still dressed in traditional costume. The women wore their *izicholo*, their wide grass-woven crimson hats, and copious decorations worked in tiny white, red, blue and black beads - necklaces, bangles, hat bands, testimony to their enduring cultural conventions. As we laboured through rural Zululand we could sense the grandeur of it all, the great curve of time and history.

We stood in silent remembrance where the battle had been fought, beneath the great rock of Isandlwana overlooking the Nqutu Plateau and the Buffalo River. It was an eerie place. The outline of that brooding mountain was like a sphinx, the mirror image of the sphinx on the collar facing badges of the 24 Regiment of Foot, added years before after they had seen service in Egypt. Here these lads from the Welsh valleys of the Rhondda, the Towy and the Taff must have stood and watched that horizon-wide Zulu army descend from the heights of Nyoni. They must have nervously fingered their Martini-Henry breech-loading rifles, this line of little over a thousand men, and watched as Cetshwayo's warriors, twenty times their number, swept up the green sward to meet them in a formation of the horns of the buffalo - sharp wide flanks and central battering head. Very few were to escape the onslaught that day.

There, right in front of the great rock, were the dongas. Burt Lancaster had gone, but his memory lingered on. We were to find hundreds of gullies all over Zululand. Then we traced them north across the White and the Black Umfolozi rivers, across the Pongola, back across Swaziland, north towards the Limpopo, on across Zimbabwe as far as the immense arc of the Zambezi. They occurred from Zululand to Zambia, over a thousand miles from south to north, and several hundred miles wide. In every one we found the organ-pipes, the stone tools, everything. We did not know the whole story as yet, how these features related to the landscape, the prehistory, the climate, the response to changes in global temperature, but we felt we had made a start. This was archaeology on a sub-continental scale.

※ ※ ※ ※

It was time to bring in another line of enquiry, another angle on the problem. It was time to look at the flowers and smell the roses, or rather, the acacias. It was time to talk to the trees. They may not look it as they stand, sturdy and immovable, but in fact trees are very sensitive things. Any slight change in the barometer for any length of time and they will die away in one area and re-establish themselves in their own distinctive habitat somewhere else. For example, during the last cold spell in Europe twenty thousand years ago, when Kent was having its worst winter for years – literally polar – the hornbeams, beech and birches of the Weald all packed their trunks and disappeared to Spain. It was too cold by far for them in Tunbridge Wells.

There is a theory to be developed here. If trees are so sensitive, and if we know what climate they will tolerate today, and what they will not, then whatever information we can recover from the distribution of trees in the past will be an indicator of the heat or cold, the dryness or rainfall, of that time long ago. It would be another way of reconstructing ancient climate. That's assuming we could find the evidence, which was in itself a big assumption. To test this theory, what we needed in Swaziland was a palaeo-botanist, a tree person, someone who knew how to read the plant life of the past, and I had just the one in mind, provided I could uproot her from gardens of Britain and transplant her in Africa.

I had worked with Jules in Gaza, and we had evolved a new procedure for the study of ancient flora. Up until then, the techniques developed to recover plant remains from the past had all been worked out in America and Britain, on the plains of Ontario or the peat bogs of County Antrim, for example. They involved the analysis of ancient pollen grains. Amazingly, in the right conditions pollen grains can survive for thousands, if not millions of years. If they can be recovered, they can be examined microscopically to find out what plants had grown in that particular place in the past. That's fine for the marshes of Canada or Ireland. The problem was that in the sub-tropics, where we were working, pollen was not preserved well, if at all. So, instead, Jules had dreamed up the idea of analysing something that we knew had survived - prehistoric charcoal from ancient hearths. By looking at the cellular structure of the burnt wood through a scanning electron microscope we hoped to be able to identify the exact plant or tree from which it came. It was achingly slow, and they had said it couldn't be done, but we had done it, and eventually had convinced others that it was a viable method.

On one of my U.K. visits, Jules and I had lunch together in South Kensington. She was inclined to be temperamental, but fiercely loyal when committed, and with a phenomenal brain. When I deemed the time was right I dropped

Africa into the conversation. It was like lighting the blue touch paper; I waited for the explosion. But in fact she was enthusiasm personified, suggesting lines of approach, learned professors to see at various Universities and which botanical gardens to contact. She wanted to start right away.

So here we were, in the bush, considering the trees. Ralph had taken Jules around Hlane and Mlawula in the same way he had shown me around two years before. She was totally captivated on the one hand, and completely intimidated on the other.

"You do realise, David," she said, in one of those 'Me scientist, you idiot' voices, "that there are over one thousand woody species in this part of Africa alone. Compare that to Europe and it's more than ten times the number. And we are going to have to collect every one, trunk wood, twig wood, leaves, flowers, fruits, the lot. It is going to take forever before we can attack the ancient charcoal, that's assuming there is any."

"Well, we'd better make a start then," I offered quietly. And with that, we did. The project was to take the best part of ten years and involve a whole crew of people, but it was the way forward.

The trees fascinated Jules. She enlisted the help of Ben Dlamini, an old Swazi from the much-neglected Herbarium in Mbabane. He knew the trees like the back of his hand, especially their traditional medicinal uses. Locally, there was a lot of money in that side of botany. Jules was fascinated by the way the Swazi healers made potions from the bush flora, and while she was in the collecting stage would tell me what each one was used for. She would show me the tree and then tell me its medicinal properties.

"You see this flat-crowned tree?" She pointed to a flat-crowned tree. "You should know this one. It's an albizia, related to the acacias."

She read from her 'bible', *'Trees of Southern Africa'* by Keith Coates Palgrave.

'The bark of the flat crown albizia, *Albizia adianthifolia*, is poisonous, yet the Zulus make a 'love charm emetic' from it, and in Mozambique it provides a remedy for bronchitis.'

I looked impressed. "So in a charm offensive you can puke sweet nothings in your girlfriend's ear and cure a chesty cold simultaneously? Great! I can see the universal appeal of that."

"Don't be so stupid!" she admonished.

There was one marvellous old tree in the Nkumbane Valley. The Swazis called it *shamfutsi*. It had a wonderfully hard russet heart wood. It also had remarkable pods, shaped like a flat oval wooden box. When the pod opened like a cigarette case it was filled with vermilion and jet seeds. Ralph called it the wooden ashtray tree. Half an empty pod made a super ashtray. I read from the 'tree bible.'

'*Afzelia quanzensis*. The pod mahogany. An infusion of the roots provides a remedy for bilharzia, a cupful of the liquid being blown into the bladder through a very thin reed. It is also used for certain eye complaints.'

Ralph had warned me about bilharzia. It is a tropical disease that was rife in Swaziland, especially in the lowveld rivers. It was identified in 1851 by a Theodore Bilharz. I looked it up in a medical text book. 'Schistosomiasis (aka Bilharzia),' it said, 'is a parasitic invasion by a self-copulating fluke in the intestine or bladder.' That's pretty acrobatic to begin with, I thought, but there was more. After masturbating away through the innards the eggs are evacuated into water, where they hatch, swim around and find another host, this time a fresh water snail, and indulge in another round of self-abuse in the snail's liver, which finally bursts under the strain of continuous onanism. New larvae are sent back to the water and enter through the human skin and the whole process starts all over again.

I didn't fancy catching that, especially the liquid blown into the bladder bit.

Jules' appreciation of the trees of Swaziland progressed. Some were very easy to identify, especially if they had fruits or flowers. But the acacias always remained a real nightmare as there were so many different species – up to thirty or more in Swaziland alone. We found different types of them, from the dry bush, all the way up to the wettest highlands. Just when you thought you had keyed one out using the 'tree book', some other feature would rule against it. We spent four whole days trying to key out one particular tree, going backwards and forwards through 'the bible', checking first one feature then another. No luck. Finally I contacted one of the game guards at Hlane.

"Do you know what this tree is?" I asked

"*Yebo*, Dokotela!" I know this one!" he replied. Great!

"So? What is it?" I pressed him.

"I have forgotten it, Dokotela!" he said with a disarmingly wonderful smile.

The walks through the bush with Jules opened my eyes to a totally different aspect of the environment. She was not just looking at individual trees, but what she called 'assemblages,' different species of trees that grew together, some competitively, some symbiotically, in each micro-climate. She taught me all the names, the scientific, the common names, the Swazi names. I remember walking across a dry watercourse when we came upon an unusual group of very rough, black barked trees with nothing growing near them. A flock of speckled guinea fowl ran scuttling between the trunks, their red caps and turquoise necks adding a touch of colour. They hurried on into the grass beyond, as though not wishing to linger beneath the branches of this particular grove.

"See this one?" she said. "This is the tambuti tree, *Spirostachys africana*. It's very toxic, strychnine or something like it. When it drops its leaves it poisons the ground. Don't ever cook with it. The smoke makes people really ill."

"This one is quite common here. It's the indaba tree, *Pappea capensis*. The 'bible' says that an infusion of the leaf from this is used as a cure for sore eyes and that venereal diseases can be treated with a preparation from the bark. There, if you have an eye for the ladies at the 'Why Not' disco, this is the stuff for you."

The 'Why Not' was an insalubrious bar and dancing establishment near Ezulwini where all the tarts hung out.

Some trees were exceptionally rare. Deep in one of the Lubombo ravines we found a solitary example of a tree called the Lubombo wattle, *Newtonia hildebrandtii*. Curiously, its heavy overlapping branches seemed to have fused together. Jules reckoned it was a surviving relic of an earlier climatic period. Other trees were very common, like the sickle bush, *Dichrostachys cinerea*. It was a small bush with extraordinary yellow and pink flowers that looked like miniature Chinese lanterns. It was sacred to the AmaSwazi. Boys had to collect branches of this tree, *lusegwane*, and bring them to the annual *incwala*. If the leaves were still green by the time they reached Lobamba, it meant the lad was pure and could attend the ceremony. It must have been tough on the ones from the borders of Swaziland.

One of the great families of trees in Swaziland is the combretum family. The common names don't really tell you a lot about them. Universally they have four-winged fruits and in Hlane we had several different species. One variety is the leadwood, *Combretum imberbe*, which the Swazis called *Imbondvo mhlope*, the white ironwood. It is a beautiful tree with silver bark and light olive green leaves and actually is found over much of southern Africa. The black heartwood is the heaviest of all the trees in Africa. We found out why some years later. When Jules was looking at the heartwood through a scanning electron microscope she noticed tiny crystals in all the cells. They looked just like peppermints in a box. This is where the tree stored all the unwanted calcium that it had drawn up from the ground water as it grew. That's why it was so heavy.

"You see," said Jules, "trees don't have any excretory organs as we do, so they have trouble getting rid of waste products. This is one way. It synthesises the excess calcium from the ground water as calcium oxylate and stores it in the dead heartwood cells."

When Jules was explaining this same feature some time later to a group of visiting Americans, one of them, from Atlantic City, New Jersey, thought for a while and then suddenly interjected:

"Oh my! Do you mean to say that this is how plants go to the bathroom?"

The combretums were numerous. There was one with huge medallion like fruits. With other varieties the fruits were small, like a penny. There was one species that rarely grew into a tree at all, preferring to scramble over other vegetation that had already made it to the canopy. This was *Combretum paniculatum*, the flame combretum. It was well named. It had masses of bright fire-red flowers all over its branches. I looked it up in the 'bible,' to see if we could grow it in the garden. To my astonishment I read: 'In African medicine the crushed root and stem scrapings are mixed with dog faeces, dried, powdered and then sprinkled on food to treat madness!'

In the various tomes we had on botany many of the trees had German names associated with them. 'So-and-so *krausi*,' or 'this-and-that *zeyheri*.'. It turns out that the Germans were very keen on collecting plants during the 19th century. But more importantly, they were besotted with classifying them. So we have names of eminent botanists like Friedrich Welwitsch and Carl Zeyher plastered all over the trees of southern Africa. I wonder if Herr Doktor Zeyher knew that his personal bush willow, *Combretum zeyheri*, was good for easing backache or curing haemorrhoids.

Jules was a wonderful collector and a great companion in the bush. Slowly she amassed a huge collection of material and identified hundreds of tree species that later would be used to sort out the charcoal fragments we hoped to excavate from some of the archaeological sites. She brought out her own PhD student too, Jane, who curiously enough was a distant relative of John Rouse Merriott Chard, aka Stanley Baker, that officer of engineers of Rorke's Drift fame. Jane was rushed about the veld by Jules with plant press, hand lens, saw and secateurs from one tree to the next. Seizing a clump of leaves Jules would begin her harangue, saying:

"Now come along Jane. What is this one? Look it up! Come on, it's so easy!"

Watty thought she might be driving Jane a little hard, so he considered he might add a species or two of his own to enliven the research. Watty was good at finding really obscure fruits or flowers in the remote areas to which he was travelling. For instance, he once came home carrying a double capsule fruit with a long branch attached to it. It looked exactly like a penis with warty testicles, and that is how he showed it to Jules. It turned out to be from a toad tree, a rare tree from the bushveld. On another occasion, when he had just come back from a trip across the border, he gave Jules a long branch with tiny leaves on, saying he had no idea what this was. Did she know? She rose to the bait.

"Now come along, Jane. We'll soon key this out. Get the book!"

They pored over it all afternoon and finally admitted defeat.

"You want to know what this is?" grinned Watty. "It's tea, that's what it is. There's a plantation across the border."

Jules was not amused. Non-indigenous plants were taboo, not to be looked at, let alone admired or worked on.

Another incident took place at Hlane when she was staying with Ralph. Ralph was inordinately fond of curry and was always cooking up 'impala Madras,' 'warthog vindaloo' and the like. He had bought all his spices from the Indian spice market in Durban, an extraordinary aromatic place reminiscent of the back streets of Bombay. Among his most recent acquisitions were two hundred and fifty grams of dried curry leaves, which incidentally is botanically speaking *Murraya koenigii* and is related to the citrus family. Not a lot of people know that, and certainly Jules didn't, mainly because this little plant comes only from Sri Lanka. In the bag with the leaves were some seeds of the same curry bush, which Ralph had planted outside the door of his house. They took a long time to germinate, but in the end two tiny leaves appeared. Ralph was thrilled; he would have fresh curry leaves for his 'wildebeest biriyani.' But he had reckoned without the 'flower power,' as Jules and Jane were called. Jules saw this strange plant, pulled at the leaves and accidentally uprooted the whole thing.

"Now come along Jane. Key this out," she hectored, until brought to a halt by a shout from within.

"Hey, that's my fucking curry bush you've just made off with!"

CHAPTER TEN

INCISED MEANDERS

If I had my time all over again I would try to design something useful rather than destructive - preferably a lawnmower.
                                       Mikhail Kalashnikov

When we were working down in the bush during those early days, before we had our own camp, we used to squat at Hlane, only a few metres from Ralph's house. Being out in the wild was always a thrill. Arriving back at camp after a long hot day, we would shower at Ralph's, then build up the fire and sit around telling stories of the wild. Ralph had a fund of tales of old Africa, about animals, explorers, wagon roads and adventurers come to grief. We were particularly fascinating by his encounters with snakes, of which there were many in the bushveld, some of them very poisonous like the Gaboon viper or Rinkhals cobra. Actually we saw hardly any snakes the whole time we were there, and certainly no one was ever bitten by one. But his stories always held us in thrall, as we sat in the velvet darkness with the flickering embers lighting our faces, imagining a brush with that flickering tongue.

One story that Ralph told was about the time he was riding his motorbike on an isolated bush track in the Lubombo. He loved to scramble on his Yamaha through these gorges, and one morning he came around a corner on a narrow track and there in front of him was a black mamba reared up to strike.

"Imagine that. His head was eye-high, and he was very cross, eh! He must have been four metres long, as thick as your arm and aggressive as hell. You know, as little as two drops of that venom can kill you. He had his narrow hood up and was just gaping at me, his dark mouth open. It was 'break out the brown trousers' time for me, for sure."

As it turned out, the standoff ended in stalemate, and the mamba finally lowered itself and slunk off into the grass. We were duly impressed and scared stiff by turns.

On moonless nights, lying in a sleeping bag on the slowly spinning earth, the whole firmament seemed to hang there, so bright and shimmering, and so close we felt we could reach up and touch it. Sometimes there would be arc after arc of shooting stars, choreographed in an erratic fiery train across the galaxies, like flying sparks from God's grindstone. The southern constellations look very different to ones in the northern hemisphere. The Milky Way is a blur of luminosity from horizon to horizon, a diaphanous stellar mist suspended across the heavens. Orbiting it are the two smudgy white clusters of the Magellanic Clouds, like two glowing lamps. They are made up of millions of stars spinning in space and trailing a tail of stellar debris and gas 160,000 light years long that has been flung out by the elemental forces of the universe. The Clouds were unknown to European astronomers until first recorded when Magellan's flotilla circled the globe in 1520.

The Milky Way passes through Carina, the Keel, named after the keel of *Argo Navis*, the mythological ship of Jason and the Argonauts. On this side of the Equator the heroes of the Golden Fleece still navigate eternally through the southern heavens. In the stern of the Argo sits Canopus, the second brightest star in the sky. And elevated above us was Crux Australis, The Southern Cross. Although the smallest of all the constellations, it has the highest concentration of bright stars of any one of them. Using the two pointer stars, Alpha and Beta Centauri, you can find due south, way down the sky. The Cross would turn during the night, and in the morning could be low, on its side near the lightening east, but the calculation with the two pointers always aimed due south.

Looking up each night, and in that pre-dawn darkness, I wondered what the people of the Stone Age had made of it all. What supernatural forces had they divined in these heavenly beacons? What pictures did they imagine in the sky amongst these pale orbs and shimmering dust clouds, their outstretched bony fingers charting the cosmos in a time before ships, before mariners? Did they also weave their own legends among the bright points of light and darkling recesses of this luminous nightly spectacle? It is said that the Bushmen could see the moons of Jupiter with the naked eye. To the Stone Age hunters the heavens must surely have been a metaphysical chronicle, steeped in myth and imbued with profound mystery.

When the moon was high and the bush was bathed in a soft silver glow, the night calls reverberated across the surface of the ground in an invisible opera of sound. Pairs of black-backed jackals howled at the sky, whining and squealing in an eerie duet. Impala snorted under the trees, warning of danger stalking the dusky veld. Far away, in the woodlands of Mlawula, we could hear the macabre whooping of spotted hyenas, the sisterhood calling their young to a scavenged kill. And as a harbinger of the dawn, fiery-necked nightjars plaintively whistled their repetitive quavering call, 'Good Lord, Deliver us! Good Lord, Deliver us!' There were so many unidentified sounds and rustles in the grass. The bush was alive with nocturnal busying in that penumbral blackness.

As the blush of rose appeared in the eastern sky, faintly outlining the ridge of the Lubombo and heralding the dawn, I could hear Colin getting out of his sleeping bag. He was making the coffee. In the still air, with only the dawn chorus to break the silence, the noise of his every movement was magnified. Minutes later he banged the enamel mug down by my sleeping bag.

"Wakey, wakey, director!"

"Colin, I heard you get up. I heard you break the twigs to kindle the fire. I heard you clattering about with the kettle and the jerry cans. I even heard you trying to open the coffee jar. How can I possibly be asleep?"

* * * *

Today we were going to investigate a particular problem we had pondered several times before and not really finalized an answer. The Lubombo mountains ran across the eastern boundary of Swaziland like a huge dam wall, extending hundreds of miles north as far as the Limpopo and south into Zululand, effectively cutting the lowveld off from the sea. Yet Swaziland's four major rivers, the Komati, Umbuluzi, the Usutu and the Ngwavuma somehow ran through this escarpment to the Indian Ocean. Since they clearly didn't climb over the mountains, or pond up against them, we wondered what had taken place here to allow this to happen?

We set out, Watty, Colin, Andrew and me, driving along the base of the escarpment and ascending the main road to Siteki, passed the headquarters of the Tit for Tat Bus Company, passed the Bamboo Inn and the Good Shepherd Hospital. The little settlement of Siteki was on the road to Mozambique. The border was only a few miles away. From the border post you could see the city of Maputo, Lourenço Marques that was, on the coast. Siteki had been developed by Europeans in the twenties and thirties as a place to escape the stifling heat of the bushveld or the malarious swamps of the coastal plains. The sign above the general store advertised fish and chips, ammunition and coffins. As the white settler farmers moved away, Siteki was to become renowned for its government school to train traditional healers and diviners, a fascinating mix of botany, spiritualism, natural science and raw superstition.

Beyond Siteki the road deteriorated to a track leading south along the crest of the Lubombo, with stunning views two thousand feet down across the bushveld alternating with the dramatic beetling cliff faces of the scarp. We had tested the rock faces of this ashy volcanic lava and found it was the hardest rock we had ever recorded. On the precipitous crags we could see clumps of the rare Lubombo cycads. These ancient plants are similar to the ones in the highveld near Ngwenya. Their fossils had been found way back in the coal measures three hundred million years ago. Their trunks stood up to fifteen feet straight out of the rock, and some had huge salmon pink cones in clusters among the fronds. There were even rarer forms of euphorbia next to them; the Swazi euphorbia, with a slender stem and cactus-like candelabra head, which only grows on these cliffs near Siteki and nowhere else on Earth. The combination of the grey and lavender rock faces and the extraordinary plants made it seem like a scene from another world. Looking east, the dip side of the Lubombo was the antithesis of the scarp. It fell away seawards in gradual shallow surfaces towards the Indian Ocean maybe fifty miles away. Small valleys cut through this lightly tilted tableland to drain the rains that fell here, and the ground had been tilled in places to grow maize.

After maybe two hours' drive along the top of the Lubombo we came to a bend in the road and pulled off to the edge of the scarp to get our bearings. Looking down into the lowveld we saw way below us the green expanse of the sugar estates at Big Bend, like a huge jade plain, and meandering through it was the Great Usutu River. Close to the sugar mill, which looked like a toy from where we stood, the river made a wide loop, hence the name of the estates, before flowing straight for the chasm which would lead it through the Lubombo and out onto the plains of Mozambique. This is what we had come to explore. We were going to descend from the heights of the Lubombo into the deep recesses of the Usutu Gorge.

Taking all our equipment with us we left the Land Rover and began to hike. At first the walking was easy, a gradual convex surface slowly edging down towards the gorge. The country was open and green with a new flush of grass where it had been burnt a couple of weeks before. A man appeared as if from nowhere and overtaking us with unusual haste disappeared over the ridge. We followed the same path. Steadily the ground became steeper, scattered with flame acacias and tree aloes. From there we could see it plunging precipitously into the depths of the gorge. Progress became slow as we negotiated the twists and turns of the rocky path, and the vegetation becoming progressively thicker. We disturbed a troop of vervet monkeys that gibbered and peered at us warily before scrambling off through the branches with consummate ease. Their light, skipping step contrasted with our heavy, halting progress.

The path became thickly overgrown with lianas and an especially troublesome acacia, the river climbing acacia, *Acacia schweinfurthii*, named by one Georg August Schweinfurth, yet another of these mittel European botanists from the 19th century. Well, he certainly picked a bummer in this one. It has long, almost invisible branches that blend into the overall colour of the bush and these are covered with tiny needle-sharp hooked spines. One by one we walked into this tangle, only to become ensnared like Gulliver by the Lilliputians. The hooks caught and pulled at our skin, snaring our legs and arms even through our shirts. Each time, one of the others had to come and extract the thorns to allow us to continue.

We could hear the river far below us, but we couldn't see it for the dense cover. Purple crested louries called in the tree-tops with their deep, resonant cackle, the sound reverberating among the crags. Onomatopoeically the Swazis called them *ligwalagwala*. When we could see them they were a technicolour mixture of orange, green and lilac, and when they flew their wings' feathers shone with an iridescent flash of metallic cerise.

Finally we made it to the bottom of the gorge and walked out onto a natural grassy terrace. Even in winter the atmosphere was stifling and airless, but the river, flowing low over the sandbanks, was a fine sight and added a cooling presence to a scene of old Africa. Huge sycamore figs, seventy feet or more high, spread their heavy branches over the river, the greenish-yellow bark blending with the green-barked fever trees along the water's edge. Away from the river was an impenetrable thicket of innumerable spiny and thorny trees and shrubs. Somewhere in these thickets, hiding from view, were narina trogons, bright green birds with scarlet breasts. We could hear them calling their distinctive hoot-hoot call in the forest. They were accompanied by the disembodied mournful whistles of tchagras and bush shrikes. One called in a monotonous series of drawn-out ghostly pipings that sounded like someone laboriously pumping up a bike tyre. We named it the 'bicycle pump bird.' Along the riverbank surprisingly well-worn paths led along the gorge as it curved round in a series of deep-set meandering arcs. We must have been almost at sea level now, the river running sluggishly from one deep pool to the next among the exposed sand banks. Much of its waters had been abstracted to irrigate the sugar upstream at Big Bend and it was flowing at a fraction of its summer flood.

We began to notice an eerie stillness in that sunless gorge, as if it had just been abandoned after a plague. As we walked we came across a group of huts, crudely constructed with bundles of reeds and quite unlike the neat grass-thatched beehive huts the Swazis made. A fire smoked listlessly in front of them and a few blackened pots and pans lay nearby but the homestead was deserted. We could hear voices speaking quietly on the other bank, which in fact was in South Africa, if national boundaries meant anything in this dark and remote defile. It was as if the people had fled across the river at our arrival. I remembered the man who had overtaken us in such an uncharacteristic hurry. Had he raised an alarm, and if so, why? We continued along the sinuous path, but at every turn we observed the same phenomenon, a cluster of reed shelters left in unseemly haste. Here, a meal had been left abandoned. There, a homestead was inhabited only by a few brown and white goats.

"Oh look, director, it's a little vegetable garden," said Colin in a confident voice as we passed one of the huts. He pointed to a bed of seedlings.

"Jesus, that's not a garden Colin, its ganja!"

We had just stumbled on another part of Swaziland where they made their living from recreational weeds. It was *insangu*, aka marijuana, the same as they were growing on an industrial scale near Piggs Peak.

We became aware of someone spying on us through the branches of the figs on the other side of the river. His face was black, much blacker than the AmaSwazi, and he had heavy features I had not seen before.

"These are not Swazis," I whispered. "They are Mozambicans! Look at his face; look at those huts."

Mozambique was at that time in the middle of a devastating civil war. We had been told about refugees fleeing the fighting, bringing Kalashnikov rifles to trade for a bowl of mealie porridge. It was rumoured that in these remote places there were rifles everywhere, hidden under floors, up in the thatch and in dead tree trunks. I had no idea what the war was about, but I had no intention of becoming a statistic in it. Who knows what contraband flowed along with the river through this gorge? Insangu? Arms? I felt a distinct twinge of unease. What did the local generalissimo of guerrilla forces care that all we had come to look at was the wiggling river. Alarmingly I wondered how many AK-47's were pointed at us right at that moment.

As soon as we could find a break, I led a tactical withdrawal into the thicket, *Acacia schweinfurthii* or no. As we legged it back up the steep slopes a great deal more speedily than we had come down, I imagined I had these sort of shooting pains in my rear end. We didn't stop until, legs and arms bleeding from the thorns, we stood in an open clearing near the summit and could see near-naked herd boys and Swazi cattle grazing on the new pasture – civilization. Regaining the Land Rover, we unpacked the lunch from Colin's back-pack.

"Do you mean to say that I carried that lunch all the way down into the gorge and all the way back up again for nothing? Well I do think that's a bit steep, director."

"Absolutely right, Colin; I thought it was a bit steep too, especially that last bit!"

We sat in relative silence eating the usual bees flies and avocados. Had we just had a narrow escape from the forces of darkness, or was I becoming paranoid? At any event, we had achieved our objective without loss of life.

When we had finished lunch, Andrew and Watty reviewed their understanding of the landscape through which we had walked. Their interpretation revolved around the work of a geology professor from Durban, who in the 1940's had noticed that if you were to bend a sheet of cardboard over the whole of southern Africa, sea to sea, no sharp peak would poke out through the top. The mountain peaks were not so much all at the same height as all congruent in the same slightly convex plane, with valleys eroded around them. It was like a praline ice-cream melting in the sun, with the cream running away and leaving the hard bits still standing to demonstrate the original shape of the brickette.

To understand why this was happening, we had to go back to the splitting up of the great super continent, Gondwana, some two hundred million years ago. After the break-up the African coast began to take on the shape it is today, with Australia and India rafting off to the east, and South America breaking off from the Cape Coast and being pushed westwards as the Atlantic Ocean opened up. During the Falklands war of the 1980's one geological wit suggested that the islands should be returned to where they really belonged, South Africa, from where they had been ripped tens of millions of years before.

At first, as Africa took shape, the high land in the middle started to erode, its surface slowly planed down into a vast, almost featureless, shallow dome. Rivers flowed across this flat continent to the seas around its perimeter, meandering from side to side in a series of incoherent squiggles and ox bows as they did so.

However, after a time, movement beneath the crust eventually forced the dome to rise and curl. The uplifted edges again became vulnerable to the elements and they began to erode again, and so, slowly, slowly, age by age, as elevation progressed, the rivers began to etch their way into the surface, meander by meander. The more the surface was uplifted, the more they cut down in their beds, perfectly mirroring their original mature flat courses. These river courses were the last vestige of an arcane landscape. That is what these deep meanders were in the Lubombo. We were looking at shadows from that ancient African surface one hundred million years ago or more.

So how did the deep trough of the lowveld, behind the Lubombo come to be there, overlooked by this dramatic scarp? It seems much of the weaker rocks of the lowveld and middleveld had been rotted to dust during a long period of tropical weathering after the break up, decomposing to great depth under the influence of water and air. I remembered seeing the lurid pink disintegrated granites on the way down to the Ngwempisi River. Watty colourfully referred to them as 'knackered granites,' until one of the volunteers, unfamiliar with the jargon, wrote in saying he had looked up the term in every geology textbook he could find and still hadn't located it. Could we give him a relevant reference for it?

Initially, this weathered bedrock stayed where it was, 'knackered' but in place. But as the edge of the continent rose, these weathered rocks of middle and lowveld had been slowly excavated away just as surely as if their soft decayed bodies had been carted off in truck-loads. The Lubombo was made of chemically very much harder rocks, and still retained part of the original profile of the uplifted dome. The rivers had had time to incise their courses into this unyielding ridge, resulting in the particular phenomenon we had witnessed today, the Usutu gorge. The rivers flowed in incised meanders. Not only that, if we were to project the un-eroded dip slope of the Lubombo from the coastal plain westwards, at the same angle, arching over the lowveld void, it would match the summits of the Mdzimba Mountains or the Ngwenya range in the highveld. That was the old Africa surface.

"We are looking at fragments of a landscape that is one hundred million, maybe getting on for two hundred million years old," said Watty. "It is unique in global geomorphology! It can help us understand the whole process of how continents form. That high terrace on the Umzimnene we saw some time ago with all those man-made choppers on it, the 'load of old cobbles,' represents only the last two million years of one hundred million years of time. It is really astonishing"

It was truly astonishing. Watty had the ability to describe these broad swathes of Earth history, recounting the aeons of time with a sweep of his hand, from the sea to the distant blue of the mountains of the edge of the continent. I gazed westwards in amazement. What a phenomenon this was, Africa!

* * * *

We drove the long road back to camp, arriving after dark. Finishing supper, we sat in the firelight, each with his own thoughts and perceptions of the day, mulling over the journey and the momentous conclusions about the Lubombo gorges. I heard Ralph's Land Rover return from a trip to town and faintly registered the doors closing. I was cradling my wine glass between my knees with both hands, watching in the glimmering flame-glow as a small beetle crawled out of the wood near the embers and attempted the arduous ascent over the toe of my boot. What a timeless place Africa was, from this small beetle to millions upon millions of years of evolution.

Just then I became aware of a pair of human feet next to mine. Someone had sat down beside me. I stared at these feet, exposed podgy toes sticking out of a pair of high-heeled satin mules, the diminutive toenails painted in day-glow orange. They were so incongruous, these painted feet, beside my beetle and my boots. I sat hypnotized by them for some seconds before looking up, and there, next to me, sat a pneumatic, blousy woman in her thirties with a ski-jump nose and a broad, florid complexion. She had dyed blond curls and was wearing a short-hemmed, low-cut dress and full make-up. She gave me a wide, toothy beam.

"Hi there! I'm Mandy. I've just come with Ralph."

She had a sort of high-pitched girlie voice, with an accent somewhere half way between breathy Birkenhead and Bowling Green, Kentucky, and a touch of Loch Lomond thrown in from time to time to confuse the unwary. Ralph appeared from the shadows.

"This is Mandy Hawk, a friend. I've been to town. I picked her up there."

She seemed so out of place in the middle of the bush, a dainty, well, almost dainty, exotic plant growing among the thorny scrub. She chirruped on.

"So you're the archaeologist. Ooh, how truly, truly exciting! Do you dig people or do they dig you? Do tell. Don't you archaeologists do it with a brush or something?"

She giggled and plunged on.

"Have you guys got any interesting tools you can show me? Know what I mean?"

Her mascara eyelashes fluttered like bats at twilight.

"And what have you dug up lately?"

"Well, nothing quite as extraordinary as Ralph seems to have dug up," I countered.

She didn't notice the allusion. I considered explaining the evolution of African landscapes, then thinking better of it, I told her about the dagga fields of the Usutu Gorge and the Kalashnikov rifles, which made her squirm excitedly. In return, Mandy confided that she was an environmental groupie who just loved to come from time to time and find her roots in the bush. She screwed up her brow knowingly, head on one side.

"It's really real, and so meaningful, isn't it? The bush and all that? And those poor animals and things! I'd hate for them to get lonely out here, know what I mean?" she said with a pout.

The following morning Mandy was nowhere to be seen. Maybe she was out communing with her orange toenails. Ralph was his usual ebullient self. The presumed rigours of the night had not daunted his spirit.

"Now, Price Williams," he began, "if this team is growing like it is, we had better think of some permanent home for SARA, don't you think? We'd better look into it."

So we did.

CHAPTER ELEVEN

UP A HILL WITH SIEVES

The society which scorns excellence in plumbing as a humble activity and tolerates shoddiness in philosophy because it is an exalted activity will have neither good plumbing nor good philosophy: neither its pipes nor its theories will hold water.

John W. Gardiner, Secretary of Health
under President Lyndon Johnson

Ralph was right. Our team was growing, not the least reason for which was the birth of our son. He arrived on March 1st, St David's Day, which for a father of Welsh descent is considered very good shooting. Needless to say he was the most intelligent boy that ever drew breath, and I felt sure he would want to be fully engaged with the expedition as soon as he was able, as I explained to him a few days later.

On the ground in Swaziland things were progressing apace. Now that the field survey, the geomorphology and the palaeobotany were well established it was time to obtain more detailed archaeological information from excavations. We had several scientific parties conducting research around the kingdom. We were also about to take on teams of international volunteers to assist with the digging. This was a system I had used some years before, inviting helpers from Britain and America to join us in our research. It was beneficial financially, as they paid the Swaziland Archaeological Research Association for their board and lodge, and for their tuition, thus contributing to the ever-needy coffers. They also provided a work force. All this meant that we would have to think about a permanent base in Swaziland.

About that time, Christoph was in house-building mode. Armed with a tome on 'How to build a thatched house', he had constructed a couple of attractive homes on the family farm. They were a design fusion of Suffolk country cottage and traditional Swazi hut, using *hyperenia* thatching grass, gum poles from the local Usutu forests, and in place of wattle and daub, concrete blocks and cement plaster. One day, when we were talking together about finding another house to rent, he said, "Davie, why don't I build you a house? You know, nothing fancy, nothing too big, nothing too expensive, just something for you and the family. You could put up army tents in the garden for the Expedition."

And that was how the SARA Study Centre came into being. Nothing too big, Christoph had said. But as we talked day by day, the project grew larger and larger. Instead of the original idea of a two bedroom, kitchen and bathroom bungalow, we added a sitting room.

"Well, I think we should make it bigger so you can hold meetings and that kind of thing."

Instead of one toilet, it extended to two.

"You don't want the others using your private one, do you? And you know something, if we extended the walls a bit that way you could fit in two extra bedrooms on the other side of the house, maybe for your visiting scientists. You could get an extra four people in there. Of course, you would have to have an additional shower, but that could go in that corner there."

So the idea grew, and grew …and grew.

"You know, Davie, since you've got all that floor space under the same roof, if you raise the walls by only a few blocks, and they don't cost that much, you could add a second storey. You would have very simple flooring, but you could add an extra two bedrooms here, and of course you would have space for another two over on that side, no, three on that side. Of course, you would have to have more toilets, and a bigger kitchen, and store rooms, and naturally you would want a separate house for the domestic staff …"

And that is how we ended up with a ten-bedroom mediaeval hall, fully-thatched, that when it was finished looked like nothing so much as a Saxon long house. It was huge. As you walked in through the front door in the middle of the building, you entered into the common room with sofas and easy chairs on one side for up to twenty five people, and a dining area on the other that could seat the same number with ease. Looking up into the roof, this central area was open to the thatch thirty feet above, supported by enormous beams reminiscent of the inside of a tithe barn. Two staircases led to the left and right upper wings of the house, and on the ground floor on one side were our own private quarters. It had endless showers and toilets, and a fully-fitted kitchen where several meals could be prepared simultaneously for at least thirty people. It was like a small hotel. We built it on a corner of land adjacent to a local private game sanctuary near Lobamba. And the great thing was that the front garden opened onto a magnificent view of the granite crags of Nyonyane to the left, and to the right was a spectacular panoramic vista of Sheba's Breasts.

Of course it had not been without its constructional problems. The increased weight of the thatch, many tons of it, all brought up from the other side of Manzini, started to bend the roof beams. This brought about a lateral thrust on the walls that threatened to burst the upper storey outwards. To counteract this, Christoph added buttresses against the outside walls, which happily also provided perfect inter-spaces for verandas front and back. He also added cross beams and uprights to brace the rafters, which gave it even more of a 'Saxon long house' look, so that's what we called it – the Long House.

But the real nightmare came with the plumbing. We had engaged an excellent plumber named Indlovu, the Elephant, but he had never in his life had to tackle anything of this magnitude. The number of 'p' traps, knees, 's' brackets, reducers, expanders, unions, olives, glands, bleeder valves, sanitary fittings, pressure pipes, shower regulators and 'T' pieces he had to screw together, without a proper plan, rather got the better of him. As a result, when the toilets and showers were finally completed, with the concrete floors laid and tiled, and we could switch the electric hot water boilers on for the first time, we found that in one wash-hand basin we had hot water coming out of the cold tap and cold coming out of the hot, another with cold water coming out of both taps, another with two hot taps, a shower that blew nothing but fetid air musically through the nozzle and a toilet that flushed with scalding steam. And there was more.

Football aficionados will recall that the 1966 World Cup was one of the most celebrated events in the history of 20th century British sport, but it also incidentally brought about one of the most celebrated events in the history of British sanitation. Picture the scene. It's the final, and England and Germany are drawn with a goal apiece. The referee's whistle blows for half time. And at that precise moment the whole population of Britain goes to the loo. In London alone ten million people rushed to the toilet and simultaneously flushed twenty million gallons of water into London's antiquated Victorian sewers which coped, just. And the final score? England 4 Germany 1. England was flushed with success.

If the Elephant had read the subsequent reports of this in the various civil engineering journals, he would perhaps have worried a little more about the scale of the Long House's own sewerage system, in the sense that a sceptic tank and French drain soak-away originally designed for four people will not necessarily handle the effluent of twenty or more, all showering and relieving themselves in the same half hour before breakfast. And so it proved. With a faint odour of the wrong kind of *eau de toilette* mixed with the whiff of shower gel, the lawn at the back at first began to heave and then to transform itself into what threatened to become an unwelcome boating lake. It was time to go back to the drawing board, or at least, to re-dig the French drain a great deal deeper.

The bore of the sewerage piping from the cesspit also had to be increased, and this too brought out the best in Indlovu, at least after the unhappy incident with the Jubilee clips. It seems that when I first bought the increased sized clips that he needed, the diameter was a couple of millimetres too small, with the result that Indlovu struggled for a number of hours up to his knees in untreated effluent. With desperation in his eyes, he appealed to me.

"Dokotela, the clips they do not work fogo!"

He seemed to be intimating that there was no way in which these clips were ever going to fit. The last word is 'Fanakalo,' a highly derived language used on the mines in South Africa and made up of words from siZulu, Afrikaans, English and bits of slang. 'Fogo' is the 'Fanakalo' word for 'nothing,' etymologically derived from the highly descriptive English, 'Fuck all!' Indlovu was telling me that the clips they would not work 'fuck all!' He was right. I got new clips. And they worked. We were all tooled up to begin. As that great poker-playing adage has it, 'A straight flush always beats a full house!' We sat in great satisfaction and gazed outside at Sheba's Breasts.

Sue has always been a keen horticulturalist and was eager to make a start on the exterior appearance of the Long House. She had visions of glades of indigenous plants, gardens bright with sinuous rills and incense-bearing trees, that sort of thing. She was a bit disconcerted when she found that the first plants that shot out of the ground like a forest after the early rains were luxurious stands of marijuana. It seems that the builders had indulged heavily in smoking joints of home-grown green stuff during the months of the construction, throwing the cigarette butts, which included the seeds, all around the outside of the house. Maybe that was how they kept their nerve when they were at the top of rickety ladders finishing the lofty gables ends, on the principle of 'Feeling high! Building high!'

The new planting began and soon the garden began to look really attractive. In the front we introduced lucky bean trees, wild pears and even a baby baobab.

This last one was not strictly indigenous to Swaziland, but the 'tree book' mentioned that 'a draught of water in which the seeds of the baobab have been soaked and stirred will acts as a protection against attack by crocodiles'. It occurred to me that if the problem with the French drain recurred, this might be a sensible addition. To hide the rear fence, which was a bit of an eyesore, we planted rows of the flame combretums, the scrambling ones with the fire-red flowers, the bark of which when mixed with dried …well, you know. Anyway, at the back we also planted vegetables that grew remarkably well, what with all the extra fertilization the soil had received.

Our carrots were just coming to maturity, with fountains of green fronds and thick roots, when out of the next-door nature reserve appeared a troop of vervet monkeys. They side-spied the carrots then descended *en masse* into the garden. These monkeys were obviously well fed; either that or they had a childish aversion to carrots, because they pulled each one up with great care, and taking only one bite out of every one left them all uneaten on the ground! We were furious, but could only watch as the large males, the ones with the gigantic turquoise testicles, rushed the troop away for another plundering raid elsewhere. After that, we built a large wire cage in which to grow our vegetables and observed with smug satisfaction the longing looks of the monkeys every time they put in an appearance.

∗ ∗ ∗ ∗

We were broadening the scope of the project and ready to welcome our first volunteers. We had promoted Toko to expedition housekeeper and installed her in her own quarters off to the side of the main Long House. We had built rooms for more domestic staff and we hired Elvina to help out. The two worked well together, and if we needed extra hands from time to time, Toko would ask her sister, the local witch-doctor's wife, to join in. On the afternoon that the first of our volunteers were due to arrive from the U.K. and USA, Watty and I picked them up from Matsapha Airport. We knew quite a lot about this group of ten recruits because they had all filled in application forms months ago. There were spaces on the forms to put information like 'health status,' 'fitness regime' and even one which asked, 'Is there anything else the Expedition leaders need to know about your health?' Reading through these the night before gave us the impression that we were about to greet a team of ultra-athletic super-people. As it turned out, what stumbled off the plane was a group of the asthmatic, the halt and the lame which would have done justice to the 'before' footage of a faith healing telethon. Still, we had time to mould them to our expedition.

Then there were the names. It is an observation I have made over the years that many female volunteers, at least from the USA, are either called Nancy or Betty, while many of the men are called Frank or Jack. That certainly made things easier for us in the fortnightly turn-arounds that we had during the digging season. It meant that we didn't constantly have to remember new names. We only had to perm the old ones to stay in the game. Coming to breakfast I would simply say 'Hi Nancy' and I could be pretty confident that someone would reply. As we got to know each group better, of course, their own individuality shone through and we were careful to make sure they all felt part of the team. To some, this journey was not only a holiday to Africa; they had intended it to be a life-enhancing experience. A number of them, especially the women, may have been through a divorce, bereavement or worse, some near-fatal illness for example. For these people, this trip might be like putting the first deliberately adventurous toe in the water of a new existence. For others the exploration into the human past in Africa was a life-long ambition, which had become a passion after reading everything about the Stone Age and the evolution of Man. With one or two exceptions, our volunteers added a new and enriching dimension to our lives in the bush. One memorable example was Eugene, from Yuba City, California.

He was in his seventies, but fit as a flea and the soul of enthusiasm and good humour. He had been an Okie, a survivor of the Oklahoma dust bowl of the 1930s. He and his dad had trekked to California in almost total penury to find work, and slowly, after his dad had died, Gene had pulled himself up until now he and his wife were comfortably retired on their small farm in the Sacramento valley, surrounded by peach orchards. He would tell stories reminiscent of Steinbeck's *'Of Mice and Men'* or the *'Grapes of Wrath'*, especially when he described the tricks his dad used when trying to force his recalcitrant male donkey to have a go at the much larger female horses to beget more mules. We asked him to tell that story several times, which afforded much mirth. But Eugene's real interests had gradually moved away from mules and had focussed on reading about Early Man, the Stone Age and the African bush. His wife had given him this trip, his first outside of the USA, to fulfil a dream, to see and feel the place where Man had originated. I remember when I handed him some stone tools to examine that had just come from the excavation, his eyes filled with tears and he looked at me, and then down at these implements in total awe. He was totally captivated; Eugene was touching the past.

He was also very willing to try anything. We once handed him a carton of the local maize beer to drink on the way home from the excavations. The cardboard carton had been rolling around in the back of the Land Rover all day and by now was warm and slightly rank, but Eugene pluckily downed the whole lot, with an expletive that didn't come from *'Oklahoma,'* the musical.

"Man, this beer smells like a rat's ass!"

Another old-timer was Jack, from Trenton, New Jersey who had degrees in all sorts of subjects and was interested in everything. But his one ambition was to excavate a stone implement in Africa. The excavation in which we were currently engaged had reached Middle Stone Age levels, and in my nightly talks to the group at the Long House I described the method these MSA people had evolved for making flaked spear points, which would have been hafted in a wooden shaft. It was a great technological advance. Over the next few days Jack religiously scraped away at the deposit in his square, minute by minute, hour by hour, barely stopping for lunch, in the hope of finding one of these points. He showed me one piece in situ.

"Is this one, David?" he asked encouragingly.

"Well, not exactly Jack. It is a shaping flake, but alas not the finished product.".

Next day the same thing happened.

"Is this what we are looking for, David?"

"Actually Jack, that's not bad. Only problem is, and don't get me wrong, it would have been beautiful, but see, its broken and this is only the tip. But excellent work Jack!"

On the last day he was working at the site before returning to Trenton and he came over to me and asked me to check out his trench for the last time. He got back in and removing his kneeling pad from the base, uncovered a large black flake.

"If this isn't one, I will eat my kneeling pad!" he exclaimed.

There, flat on the excavated surface was one of the most perfectly flaked stone projectile points I had ever seen, about four inches long, beautifully manufactured from a piece of slightly translucent black chert.

"Jack, that's the one!" I shouted.

Everyone rushed over to see, and when it came time to remove the flake from the deposit, Jack held it in his hand while everyone took photos. I allowed Jack to mark up the reverse side of the point, with the site number, the date and the square and level number. Afterwards, I took the marking pen and, dipping it in the Indian ink, wrote in brackets after the catalogue numbers, 'Jack's Point,' and to this day it is still on display in the Swaziland National Museum. Jack returned to New Jersey enraptured.

* * * *

The site we were excavating at the time was in the highveld beyond Mbabane, on the edge of a most attractive hidden valley high among the granite tors of Sibebe, above the confines of Pine Valley though there wasn't a pine in sight, only indigenous African savannah. Sibebe Hill itself was an immense exfoliated granite mountain over one thousand feet high, colossal grey-black peels of rock like the sides of a gigantic beached whale. Looking up at the smooth convex surface of the rock, its sides streaked in black and cream, one of the volunteers wondered whether this was where the All-Swaziland Skate Board championships were held! On the days we were excavating there, we would drive to the base of the mountain each morning, then hike to the site up the face of the rock through a landscape strewn with boulders and punctuated with clumps of tough baboon grass and red-flowering bottle-brush trees. Cresting the ridge we walked into a wide, grassy alpine meadow at the far side of which was a mega-cairn of giants' marbles, boulders the size of small office blocks piled naturally on top of one another and strewn across the mountain beyond. Beneath one of these was Sibebe rock shelter, fifteen metres high, thirty metres wide and curving concavely from ceiling to the ground. The earth was dry and dusty underfoot. It was the haunt of cattle during the rains, but around the perimeter of the dry area, where the rain dripped and ran off the 'roof' of the boulder, the grass was a mass of tiny splinters of clear quartz, like shards of broken bottle glass. These are what first brought the site to our notice. These were stone tools.

We laid out a two metre square grid over the flat surface of the floor of the shelter so that we could control the excavations, to record where things were found both horizontally and vertically. The squares were given co-ordinates, by letters in one direction and by numbers in the other. I selected one line of squares to begin with, six in all, so that we would be able to get a profile of the different layers through time. The assumption with all rock shelters is that they will have been the temporary stopping place for successive cultures in successive periods and that therefore the evidence will have accumulated as they brought the debris of their lives into the shelter with them – bedding materials, tree boughs for fire wood, quartz and other stones for manufacturing the implements they needed, and food such as animal bones and plants. Despite the ephemeral nature of these sporadic occupations, a few days here, a few days there, over the tens, the hundreds, the thousands of years, layer by layer, millimetre by millimetre, generation by generation, the cave floor will have built up a layer cake of the past, earliest at the bottom, the latest at the top. And since the climates and the cultures changed and developed through time, these layers should actually be visible in the changing colour and texture of each of the minute layers of that cake.

The concept of an excavation is to reveal the sequence of events that has been haphazardly left here through the ages, the layers that form the only and unwritten last will and testament of our prehistoric ancestors, who ate and slept, gathered and hunted, made love and played games, danced and sang and ultimately lived and died in these mountains. Of course it's an imperfect science. No sound, either lucid or comprehensible, has ever breached the silence of this speechless past. Yet these folk, although they remain now as mute as the stones they lived amongst, are the fragile link between ourselves and the very first stirrings of the human mind, that first footfall of our ancestors on this great continent more than two million years ago. Without knowing about these people at Sibebe and elsewhere, we might as well say that we sprang unannounced from the insensate earth only a few hundred years or at most a few millennia ago.

It is sobering to think that more than ninety nine per cent of human evolution is locked away in this chest of prehistory, which means that far less than one per cent of our whole background is actually in the realm of history. If I were to suggest that we should make a judgement about who we are and where we are heading using only one quarter of one percent of the evidence, I would be laughed out of court. That is why the stones at Sibebe are important. That is why Jack and Eugene, Nancy and Betty, Frank and I and all the rest of us were up that granite mountain, digging trenches in the dusty ground, sieving the dirt and picking out slivers of broken quartz, because imperfect or not, it is the only way to redress the abyss of our own ignorance. It's the only evidence we've got beyond history that we ever existed at all.

* * * *

"Hey *Bahbe*, Glenda from Swaziland Broadcasting wants to talk to you about what you are doing at Sibebe, and why you are here in the Kingdom," Toko reported. She always called me '*Bahbe*,' the siSwati for 'father'.

Glenda had left a message at the Long House. She had heard about us through her friends, the posh diners of the Mbabane social circuit. Was I prepared to give an interview for her radio listeners? I thought it would be a good thing to let people know what we were up to, and a splendid opportunity to illustrate the importance Swaziland has on the world archaeological stage. We met the next day on the side of a road in misty highveld weather in Glenda's outside broadcast radio.

"So, Doctor Price Williams, may I call you David? So, David, what exactly brought you to Swaziland? What do we have to offer you, as an archaeologist, that you can't find in the U.K. or wherever?"

"Glenda, Hello! What a most beautiful country you and the listeners live in!"

Actually at that precise moment, looking through the side window of the van, all I could see was unrelenting fog.

"Well, Glenda, as you know, Swaziland is made up of several quite distinct environmental areas, the highveld, middleveld …"

I chatted enthusiastically about archaeology, the palaeolithic, the mesolithic and as many o'lithics as I could think of. Glenda, not unnaturally, tried to get me to simplify the technical jargon. We ranged at speed over the Stone Age and the Iron Age and I managed to create a link, however contrived, between the living and vibrant world of Swaziland and a trench-full of gravel up a hill outside Mbabane.

"Thank you so much, David, for coming to tell us all about this interesting work that you are doing. I'm sure all our listeners will be totally fascinated by the whole approach and we wish you well for the next few years. Thank you again!"

She phoned the Long House later to say the broadcast would be on the following evening. The next night we were crowded around our radio after supper when Glenda began.

"What do we have to offer you, as an archaeologist, that you can't find in U.K. or wherever?"

"Well, Glenda, as you know, Swaziland is made up of several quite distinct environmental areas, the highveld, middleveld and lowveld, and we felt that …"

Everyone erupted in roars of laughter. The 'we felt'? Where was that? In the Long House, the four and a half minutes of the rest of the broadcast disappeared beneath a hail of jibes, jokes and general banter at my expense. The family has never let me forget that moment of enthusiastic nonsense. A week later even my own team was still ragging me about it. It was on a morning when a cold front had crept up the coast from the Cape, blanketing the whole of Swaziland in chilly mist and unseasonable rain.

"Well," I announced at breakfast, "we really can't get up to Sibebe today. It will be raining hard in the highveld. What do you think we should do? We could find something local to do here in the middleveld, or perhaps do some survey work down in the lowveld?"

"Maybe we should try the 'we felt'!" quipped one comedian.

"No! No! We can't do that!" replied Watty. "It's always pissing with rain in the 'Wee Veld!'"

CHAPTER TWELVE

A WHOLE LOT OF SCRAPING GOING ON

Everything is funny so long as it happens to somebody else.
Will Rogers

The fine winter weather returned to Sibebe and the sky was once more a cloudless sapphire. As we walked up the slopes of the mountain, red-winged starlings whistled from the rocks and a pair of white-necked ravens wheeled overhead before plunging together down the cliff face in a tumbling aerobatic display. I had invited some students from the University of Swaziland to join us so that I could explain the meaning of the Stone Age to them, and how we used the excavated evidence to find out more about it. One or two of the Swazi girls were a little on the plump side and had made rather heavy weather of the climb. Our older volunteers manfully helped them up. We arrived at the rock shelter and when they had all regained their breath, I began to describe the excavations and how the stone tools gave us an insight into the past. I jumped into the trench and borrowing a trowel scraped at the side of the square to show the different coloured layers, each written up with a white label, that marked the passage of time downwards and backwards through the millennia. Afterwards, we went over to the sieves and, rubbing my fingers through the spoil, I picked out suitable fragments of quartz to illustrate the kind of evidence we were looking for. It was my standard 'introduction to prehistory' lecture.

"And so that is how we discover our past. That is how we communicate with the spirit of our own ancestry through the mists of prehistory."

I paused for the usual reaction. None came. After a nervous few seconds I asked if there were any questions. One girl who had been watching me with growing interest throughout this peroration put her hand up. She was a young and attractive first year student with the beautiful smile of innocence Swazi girls display.

"*Hawu*, Dokotela!" she exclaimed. "You know all these things. So many things you have told us today we did not know. How many years did it take for you to learn them all at the university?"

I beamed indulgently.

"Ten years," I said. I thought she looked suitably impressed, so I continued.

"Yes, ten years. You know, I did an initial degree, then went on to read for my Master's and finally four years for my PhD. There you are."

"*Hawu*!" She shook her head, her eyes wide with surprise. "*Hawu* Dokotela! Ten years. It is a long time, a very long time. So if you are so clever after this very long time, how is it that you still have to dig holes in the ground and roll around in the mud? Why don't you get labourers to dig your holes for you? Look! Your hands are all dirty. Is this what education is for?"

We had already amassed a fine record of the last thirty to forty thousand years of the Stone Age, including fascinating new material about the Late Stone Age people some eight thousand years ago. At that period, as the present global weather system was beginning to establish itself after a long epoch of coldness, the hunter gatherers of Sibebe were crafting beautiful miniature scrapers, no more than two centimetres long, from the clear quartz crystals that they found in the crevices of the nearby granite. The scrapers were superbly finished, sharpened by the removal of tiny micro-flakes that could only really be appreciated today by looking at the working edge with a x10 hand lens. Stuck in a wooden handle with tree resin, they had been used to clean the skins of the animals they had hunted here, like grey duiker, klipspringer and rhebok. The inner surface of the pelt had been cleaned in preparation for leather tanning, a process which also required a profound understanding of the various plant oils and wild preservatives that would have collected on the rocky slopes and down in the river valley at the bottom of the mountain. The river had also been their water supply. From the finished leather they would have made bags, quivers, clothing, blankets, thongs for nets and any other equipment they might need. By this process we were trying to see behind and beyond the stone tools, to reconstruct a broader view of the life style of the age from an extrapolation of the scrapers themselves. The scrapers were the key.

"A whole lot of scraping goin' on, eh Doc?" as one of the younger volunteers so aptly put it!

Not to be outdone, I answered. "Yes, they were certainly able to scrape a living quite well here!"

Puns, jokes and personal stories were part of the humour of the day and helped to relieve the tedium of digging or working at the sieves.

* * * *

Punsters, cardsharps, Coca Cola franchisees, doctors, lawyers, housewives, writers, sculptors, women of a certain age, men of a certain type – we saw them all. It was always sad, well, nearly always sad, to say goodbye to each team of volunteers at the end of every fortnight. Being in such close proximity at the Long House, sharing mealtimes, grinding up the Malagwane in the Land Rovers every day and excavating at Sibebe forged indelible memories and a great feeling of camaraderie. At the airport, there were lots of promises to write, to send pictures, to keep in touch. We were invited to so many obscure parts of America - to Cedar Rapids Iowa, or Sedalia, Missouri or Pittsville, Ohio – we could easily have spent the next twelve months bumming our way across the U.S.

When they had gone, the whole process began again, working out who the next team would be, what they would be like, who would stay with who in which room. An interesting point that we all noticed was that so many of our USA participants had first and last names that either began with the same letter or sounded the same. Maybe it was something that was common to babies born in the 1940's. Whatever the case, we attracted loads of people with names like Amy Ansely, Betty Boniface, Fiona Feist, Lynn Flynn, Mary Morrison, Nancy Nackerlacker, Renata Russell and Sonya Stoltzenburg.

One of the most awkward moments I remember in this respect concerned a lovely lady from Ponchatoula, Louisiana, rejoicing in the name of Kelly Kuntz. Mercifully the rest of the team was a serious group of retired people and we managed to get through two weeks without a single comment. It was she herself who brought it up, on her last night at the Long House.

"Well, Kelly", I said, over drinks before dinner, "It's been really wonderful having you with us. Thank you so much for all you've done."

"Oh, David, it's been a real pleasure, you know. I didn't know what to expect, coming all this way and all, but you all have made me so welcome. I've had the best time. You want to know something? I feel I know you so well after all we've been through. But I am really not looking forward to the flight tomorrow. I always have problems, you know?"

"What kind of problems?" I prompted.

"I'll tell you. It's my name. I am not real happy with it, my last name that is."

"Oh, why is that then?" I said. What else was I supposed to say?

"Well, you know, people look at me so funny when they see my last name. It even happened to me on the way over here from the U.S.. I was checking in at New Orleans and after I had my boarding pass and all, they called me back over the p/a system. 'Would Mrs Kelly…err. Would Mrs Kelly…uh. Well, anyway, would Mrs Kelly, you know who you are, please return to the Delta Airlines desk.' So I go back and this guy says. 'Hey listen lady, I'm not even gonna try and pronounce your last name, OK?' Then he hands me my purse I left on the counter, and says, 'You know something, you really should change that name of yours. That'll help everybody a lot. Now you all have a nice day, why don't you.' What kind of a life is that David?"

So I said, "But it's a fine Scandinavian name isn't it? And wasn't there a famous pianist called Charlie Kuntz, made jazz records in the thirties or something? Still, if you're unhappy with it, maybe you should change it. Or maybe next time someone makes a remark about it or gives you the eye, why not say, 'Yeah, I know. It's my name isn't it? Well, I'm going to change it anyway.' They are bound to ask 'Oh what to?' and that's when you say 'Sheila Kuntz'! That way you get the last laugh. How about that?"

She laughed.

"Or maybe you can go back to using your maiden name? What was your maiden name by the way?"

"Schitz," she replied levelly. "So by your reckoning I'd be Sheilah Schitz, wouldn't I? Maybe I'll quit while I'm ahead and keep it as it is. You're right, I like Kuntz!"

* * * *

The week we moved into the Long House, we had been given a pair of coal-black kittens by the two gays down the road. They were an ill-matched couple. The one was diminutive, like a sort of pixie, with a turned-up nose and button eyes who spoke in an effete South African accent. He had been a ladies' hairdresser in Johannesburg and was given to wandering around outside virtually stark naked, in bare feet and a thong. The other was an imposing, portly man, with heavy jowls and a voice like a baritone singing the lead role of Florestan in *Fidelio*. He had once been an actor, he told us, in an actor-like voice. I once heard him castigating their Swazi maid, who as far as I know didn't speak any English. She had obviously been at his whiskey cabinet again. In a booming voice he shouted at her, 'And if you should so much as contrive to do that again, I will lambaste you, so I shall.' He sounded like a version of Falstaff on a bad night. They kept Persian cats and the one having just had kittens, they were anxious to find a home for them. With two young children I thought this was no bad idea.

"You will look after them, won't you?" said the Pixie. "They are pretty wild, Persians are, but they are very good with snakes!"

I wondered if they were very good with children, too, until he corrected my misapprehension and explained that the mother and father of these two kittens had killed off all the snakes in their garden. By the time we came to excavate Sibebe, the kittens had grown into large and imposing animals. And the Pixie had been right. They were relatively wild and they did kill off all the snakes. At least, we didn't see any in the garden, so it must have been true. They were very independent and came and went in and out through the windows of the Long House at any time of the day or night.

One Sunday, we had been to Manzini to pick up a new group of recruits just arrived at Matsapha Airport. They were an interesting bunch who admitted to being nervous in case we didn't recognise each other and they would be stranded in the middle of Africa not knowing a soul. As they were all wearing designer bush jackets (the ones with endless pockets), natty bush hats with feathers, huge climbing boots with three pairs of socks and carrying lavish photographic gear and khaki back packs, as opposed to the other seven local people on the 'plane who disembarked in 'Save the Rhino' T shirts and shorts, there was no contest really. Our new team had stuck out like sequinned striptease dancers in a Southern Baptist Seminary.

Sandy Stinson was an unusually attractive and superbly coiffured ash-blond lady in her thirties. She was particularly nervous and as luck would have it had lost her luggage *en route*. Matakhosa, the luggage clerk, told me quietly that instead of being routed from Washington Dulles (DCA) via Heathrow (LHR) to Matsapha (MTS), the final destination had been interlined as MAN. Someone had mistaken Manzini for Manchester, so the bag had gone to the north of England.

We held our usual briefing that evening and welcomed the group to Swaziland. One or two had brought duty free's with them on their travels and so it was party time, what with that and the *'vinho collapsico'* on offer. I made sure I spoke to Sandy Stinson who told me she was a senior accountant for Bank of America in Washington and had led a fairly sheltered existence until her mother had died last year and left her a small legacy. She had decided to travel, something she had not been able to do until now because of her mother and the house. She said she had been so nervous about flying to Africa and had already visualized a whole horror film of dreadful events befalling her – she'd be eaten by lions, bitten by cobras, attacked by elephants. In her imagination she seemed to have been assaulted by a whole African menagerie. I reassured her that this was not likely to happen, not at Sibebe anyway. The only things she might see in the garden would be the famous carrot-eating monkeys or the odd baboon. She was much relieved.

About two thirty in the morning piercing screams came from the common room. There in the middle of the room, screaming at the top of her voice, was Sandy Stinson, unclothed apart from a small hand towel over her vitals, she of course having mislaid her nightdress in the missing luggage. Everyone had been roused by the noise and they all now stood peering over the balconies from the upstairs bedrooms.

"I've been raped!" she shouted, "I've been raped. This horrible hairy black baboon came in through the window and jumped on my bed. He put his hands on my body. It was awful, truly awful!"

I eventually realised what had happened. She had opened her window to get some air while she slept and one of the cats had leapt onto the window sill and down onto her bed. I fetched the cat and showed it to her. As she was very fond of cats, the damage was soon put right and we all went back to bed. At breakfast the next morning, it was laughed off as a bad dream. In any case, Sandy Stinson's pussy was back into the snakes by then.

* * * *

We were down over a metre deep at Sibebe, and the deposit continued. It was about that time that we reached the main Middle Stone Age level, the one which produced 'Jack's point.' The MSA hunters had obviously ranged further afield because they made their larger implements not from quartz but from some of the hard quartzites and cherts from the Komati Valley, maybe thirty to forty miles away over the mountains. Proof that they were travelling great distances, or at least trading over them, came from the fact that in the MSA levels there was iron ore pigment from Ngwenya. I wondered what other events in their lives we couldn't recover. That is one of the most tantalizing things in prehistoric research. There is only so far we can go, and the rest is barred to us forever. Whatever songs or stories or spirituality these people expressed, they have left no trace and we are compelled to try to reconstruct their lives from the slivers of stone that survive. British and American prehistorians take a slightly different approach in this respect. The Brits are usually not keen to press further with an interpretation than the evidence will allow, which in some cases is not very far. They will substitute measurement for speculation – not so much how did these people live, but how long are these points, how thick are those scrapers – attempting to squeeze some extra significance out of the figures. The Americans on the other hand are remarkable for creating models, styles of human behaviour to interpret the past. With so much anthropological work being done observing obscure tribes in Polynesia or, nearer to home, paleo-Indian ethno-archaeology, scholars there feel that by studying the artefact assemblages of known groups doing known things, these comparisons can be used to retrieve to some degree the lost record of the past. So we hear about 'site scheduling,' that is, 'with these tools these people must have been doing this or that activity.' I happen to think it can add a great deal to our understanding in what would otherwise be a very sterile study, provided it is not pushed too far or that absolutely everything is related to the behaviour of some little-known tribe in Borneo.

Slowly the Sibebe deposit began to peter out and we came down to the raw red gravel of the weathered granite bedrock. I had opened a few more squares towards the edges of the overhang to check the breadth of the archaeological material. In the far corner we had gone down well over a metre in one double square, measuring two by one metres, just as a test trench. It was very difficult to work at that depth, doubled up, but we always had people who would volunteer to do the job, at least for an hour or so at a time. However, that too brought its problems. One morning we arrived at the site and somehow a cow had contrived during the night to reverse itself exactly into this slit trench. It was bellowing pathetically and its horns and head kept rising and falling above the ground. It was a most singular and unnerving sight, to see this disembodied beast rising like the Minotaur out of the middle of the earth. We had to cut a ramp down into the square, and with ropes over its horns eventually managed to haul it out. Someone suggested that it was a lucky animal; had the Stone Age people been around they would have made short work of this stroke of good fortune but there were no cattle in Africa at any time during the Stone Age, and certainly no domestic animals. Cattle had been introduced into Southern Africa by the 'Bantu' speaking people less than two thousand years ago. Before that, the local hunter-gatherers, the

San Bushmen, had lived in a wild paradise, unconcerned about ownership, or hierarchies, or working the land. They had lived in a Garden of Eden, remnants of which we could still imagine.

* * * *

It was on one of the last teams to work at Sibebe that we encountered Billy and his son Zak. Zak had just finished in high school and Billy had promised his son that if he graduated well, he would take him on a trip anywhere in the world. Zak chose Africa. After their stay with us, they were heading off to South Africa to watch big game, but first, Zak wanted to learn about the Stone Age and human origins, so here they were. Billy was not the most effective of helpers since he spent his time reeling off anecdotes about his own life. He was a short, dapper character in a natty jacket and tooled cowboy boots, the ones that have silver toecaps that turn up at the end – full of western promise. He had a perfectly clipped moustache and an elegant head of very well-kept hair, greying at the temples. As he spoke his small eyes darted back and forth which gave the impression that he was constantly on the lookout for someone coming to serve him with a subpoena. His son was different, and must have inherited his dark brooding eyes and laid back style from his mother. Zak in particular learned a great deal from the whole experience.

The excavation was coming to an end and I wanted to give the team a sense of place, an overview of where the site was, and what sort of opportunities there were geographically for the Stone Age people of Sibebe to make a living. So we took a long walk, scrambling over the boulders at the back of the site. The view from the summit was tremendous. All the mountains of the highveld were visible as far as Ngwenya, on the skyline. From one spot you could see right down the hidden valley to a small rain-filled pool at the very far end. It would indeed have been a Garden of Eden, with animals coming to drink, with plenty of berries and fruits, with clear water in the stream at the bottom of the mountain. These Stone Age people had everything they needed, provided the landscape was as it was today, and that was a very big question to which we were still trying to find the answer. I explained the view to everyone.

"Lots of prehistoric rear ends must have rested on this rock over the millennia," I pointed out. These lookout points always gave me an eerie sense of déjà vu. There were many places in Africa where I have felt the same sense of fundamental communication with the past.

Beyond the pool was a small thicket of indigenous forest, and I suggested we took a walk and had a look at what trees were growing there, and how they might have been useful to our prehistoric forebears. We descended into the valley and walked to the pool. The high crags were reflected in the water like an inverted mirror. We walked on into the trees. I pointed out the thorny *Carissa*, the ones the Swazis called '*Num Num*.' At this time of the year they were covered with sweet red berries. I handed them around. Zak reported that they tasted great, but Billy was not so sure and refused the offer. I explained why so many of the trees had developed thorns, against the browsing pressure of herds of giraffe, elephant and kudu. Zak was fascinated. He was learning some of the secrets of the African wild, just as he had hoped he would.

It was time to be getting back, so ducking under the spreading acacia boughs we started off out of the forest towards the pool. Suddenly there was a cry and a commotion. A small bird's nest was hanging in one of the trees and Billy stood there bald as a snooker ball, in a state of utter confusion, all dignity scraped away. Impaled on the viciously long spines of an acacia was his toupée. His son took one look at his hairless dad and laughing gave him a big hug. Billy looked horrified, then embarrassed, then began a smile which ended in a guffaw. Father and son had bonded. We had achieved more in our expedition than an understanding of the past. Maybe we had contributed a little bit to our understanding of the present as well.

* * * *

With the ending of the excavations at Sibebe, much of our work in the highveld was done. We reported to Johnny at one of our regular dinners at 'The Place of Winds' and I described the sequence of the Stone Age cultures we had discovered. I was particularly enthusiastic about the Middle Stone Age and its beautiful stone points.

"What date do you think they are?" asked Johnny.

"Well, I can't really say yet," I replied. "There is not really enough charcoal for a radio carbon date, and Jules will need what there is for the floral analysis. But if you compare this material with excavations in the Cape, or from Natal, I would say it goes back up to fifty thousand years or more, maybe much more, beyond the range of radio carbon dating anyway."

So much of my life has been spent worrying about the dating of the African Middle Stone Age. It was then, and it continues to be to this day - a difficult problem, still not fully resolved.

Ralph was pouring another comforting glass of the ice cold Neethlingshof Cuvée Blanche.

"Listen, Price Williams, now that you've finished buggering about in the mountains, isn't it time you came and worked in the real Africa, down in the bush? That's where things really happened, you know. There are so many sites there. And the veg and all is so much more African than in these highveld sites."

Now that the Long House was running well, how would we manage in the bushveld? There were no facilities we could use that I knew of. That is when Ralph and Johnny told me of a superb camp site they had just identified at Mlawula.

"You see, David," said Johnny, "that whole area has just been declared a national park and therefore comes under the National Trust Commission, of which you are a part. That's the place to be, old chap. Mlawula."

"Well, Price Williams, you'd better come and take a look hadn't you," added Ralph.

So we did!

CHAPTER THIRTEEN

MLAWULA DREAMS

Single-mindedness is all very well in baboons; But in an animal claiming to belong to the same species as Shakespeare it's simply disgraceful.

Aldous Huxley

There is an ancient grandeur about Mlawula, a glimpse of primal vistas which transcends time and which makes it, for me at least, like nowhere else in Africa. It has probably to do with the muted colours of the rocks, dark and brooding, with the stark shape of the cliffs and sweep of the mountains, and with the rivers - the Mbuluzi, Mlawula, the Siphiso and the Nkumbane. The juxtaposition of constantly-changing elevations leads you to see it now as a basin, now as a ridge, now as a valley, now as a plain. It sits amid the secret folds of the Lubombo range, concealed from immediate view, escaping the interfering gaze of modernity. When you cross the threshold from the outside, it's as though you pass through an unseen door into an undiscovered world. Yet this is quintessential Africa. From the top of the escarpment the landscape captures the mood of the Crater Highlands or the Great Rift Valley of East Africa, though by comparison Mlawula is a pygmy. I asked a Swazi traditionalist what it meant, Mlawula, and he said:

"Dokotela, Mlawula, it means … it means … Mlawula, that's what it means!"

Maybe he was right. Maybe Mlawula is untranslatable. It is what it is, incomparable, like nowhere else I have ever been. If I was in love with Swaziland, I was to become totally captivated by Mlawula.

The shape of Mlawula had been ordained some two hundred million years ago, when the great super-continent, Gondwana, began to break up. A huge rent, a series of jagged tears, appeared along what was soon to become the eastern side of Africa, joined at the time to Australasia and India, and through these volcanic fissures was ejected hot, seething, basaltic lava, suppurating over the landscape and solidifying into a black, rocky mass. As it did so, and as the new continental components drifted apart, the emerging circumference of Africa became entangled in the process. Its new-born rim was dragged down into the crust, a geological addition which changed the chemistry of the magma. The lava became stickier, less inclined to flow, so that it clogged the subterranean plutonic chambers. Immense pressures built up in the mantle until the vents exploded in a series of catastrophic spasms, sending plumes of flaming gas and billions of tons of blistering rock and dust into the atmosphere. As it settled, flowing white-hot over the underlying basalts, it hardened to become rhyolite, a rock that is the product of some of the most violent incidents in Earth history.

Today that rhyolite is the Lubombo escarpment. Tough and unyielding, it forms the high ground. The basalts, which are chemically weak and easily rotted by air and water, have been etched out to form the valleys. All that time ago, at the birth pangs of Africa, the basalts and the rhyolites alternated about a long north-south axis, so that the Lubombo is not one ridge but several. In front of the main mountain mass is a smaller ridge, the little Lubombo, and it is this ridge that locks Mlawula away. It is on the crest of the little Lubombo that Ralph and Johnny found the best safari campsite in the world, the one that became the SARA camp.

It was Ralph who first found it, making a reconnaissance of the whole area for the Trust Commission. He walked up the rocky slopes at the back of the little Lubombo, and breasting a ridge, he stood on a high, narrow shelf. In front of him was a vertical drop down to the Mlawula River. From the crest he looked west right across the bushveld, across the broad, shallow valley of the Mbuluzi, to the distant hills of Mliba and Balegane and to the far horizon, the misty peaks of the Great Escarpment. It was majestic in its sweep of country and geologic time. Standing on the youngest rocks and looking at the oldest some seventy miles away, more than three quarters of planetary history unfolded at one glance. Three hundred feet below, along the base of the cliff, the reedy Mlawula River ran in a straight course in front of the little Lubombo. Turning to look to the east, the Nkumbane Valley lay in a parallel crease of the mountains, overlooked by the crags of the main Lubombo. Not only was it a wonderful position to make camp; there was a bonus. Many years before some forlorn hopeful had intended to run cattle in the valley and had built, right on the top of the ridge a few hundred meters away, a large circular concrete cistern which could be filled by pump from the river. The site had water.

Strictly speaking, 'Mlawula' refers to the river, a tributary of one of the four great rivers of Swaziland, the Mbuluzi. But its confluence with the Mbuluzi has given its name to the region, to the campsite and to the reserve. As for the Mbuluzi, I had first crossed it when I had initially come to Swaziland, a series of youthful streams splashing down from the sides of the Ngwenya range at more than five thousand feet. It had matured since then, growing in confidence and direction, flowing larger, lower and slower until now it was a wide sluggish watercourse about to meander its way into the Lubombo through the Mbuluzi gorge, only to broaden out onto the coastal plains of Mozambique like its sister river the Usutu further south. Just ahead of entering the gorge it describes an almost circular arc, the Loop, a huge incised ox bow between the little Lubombo and the main slope of the range, before being swallowed up in the sunless maw of the ravine. The Mlawula joins it as it makes that shadowy, sinuous journey. So does the Nkumbane with its tributary, the Siphiso. Towering above its passage through the gorge are the heights of the sacred Gwalaweni forest and the two thousand foot ridge of the Ndzindza plateau, the back slope of the Lubombo that falls slowly and softly away towards the Indian Ocean.

"Right, Price Williams," it was Ralph speaking. "What do you think of it?"

I thought it was magical, the whole place. We had driven up the concealed valley of the Nkumbane and into the Siphiso stream bed. This is where I had been shown the microliths and the 'Christmas boot flakes' on my first visit that New Year's Day. Along the shallow pediments on the floor of the valley the knobthorns were the largest I had seen. Between them the grassland flourished. There was natural game here – impala in leaping showers, jackal, hyena and herds of zebra and wildebeest. Among the wooded thickets stalked kudu, with their stately spiralled horns, the shy bushbuck, common grey duiker and the timid red duiker. Waterbuck browsed along the river, and on the plateau there were oribi and rhebuck. Deep within the rocky canyons we found klipspringer and the spoor of leopard. Down the gorge crocodile basked on the sand banks and hippo snorted in the deep pools, while troops of chacma baboons and vervet monkeys played in the branches of the ancient woodlands and along the rocky precipices. Soon we were to introduce giraffe, ostrich, cheetah and rhino, but until the development of an infrastructure of gateways, roads, and trails, it would not be open to the public. It was to be our own private preserve, our own sanctuary, for the foreseeable future. And so it was. Yes, I thought, it was truly magical. And in this regard I instinctively knew that we were in accord with our Stone Age ancestors. Mlawula was going to be where we deciphered the next part of Swaziland's prehistoric code.

* * * *

"This is the main dining area," I was saying, "and this is where we do the cooking, in these monster iron pots."

A new team of volunteers had just arrived to help with the archaeological investigation and I was taking them around the newly-completed SARA camp. We had built a waist-high hob from concrete blocks lined with refractory bricks where we could lay out a bed of hot coals from the fire. Part could be covered with a thick griddle to grill meat, and the other part could take a row of large three-legged cast iron pots that made cooking outdoors in Africa unlike any other.

The African three-legged pot is a legend in the bush, a globular, lidded iron cauldron that comes in all sizes. Locals use the small ones to boil their daily porridge, but the really large ones are used on special occasions, to boil missionaries for example. The Afrikaners call this kind of pot a 'potjie.' They have used them since the days of the Great Trek and, can you believe it, have now actually elevated the three-legged pot to the status of a nationwide sport. On a Sunday in the Republic, down at the coast or in a municipal park on the Rand, they will organise a 'potjie-kos kompatisie,' a three-legged pot cooking contest. The idea is that each team in the competition, and they are usually all men, brings along a potjie and the secret ingredients for their own jealously guarded recipe for a stew. After the cooking has begun, there is a period of quiet reflection, during which prodigious quantities of beer are consumed which apparently helps the flavours to develop their full potential. Furtive visits are made to the potjie to add clandestine herbs and spices to the meat and vegetables. After several hours, when the competitors and the judges no longer know what day of the week it is, someone has to taste all the entries, but by the time the verdict is delivered, everyone is paralytic and no one is in any state to care about the result.

Back at the camp I showed people to their quarters. The bush had been cleared in the middle of the rock shelf and caravans and tents arranged around the perimeter. In the centre of the clearing was the open fireplace surrounded by chairs. Off to the side was the hob. We would cook for thirty people or more on that hob. On the opposite side of the ridge, overlooking the Nkumbane Valley and the lofty cliffs of the Lubombo, we had built two toilets with front sides open to the panorama. They were known collectively as the 'loo with the view.' Each had a flush toilet. You could sit there of a morning and watch the ants in a nearby dead tree, or contemplate the splendour of the Lubombo range and the breakup of the supercontinent Gondwana. In the one we installed a wash hand basin, but in the other we plumbed in a bidet at basin height. Since most of our guests had never experienced one before, it gave rise to interesting discussions.

But our *piece de resistance* was the *al fresco* bathroom. On the edge of the cliff overlooking the river and the bush beyond we had built a rhyolite-flagged terrace on which we placed six old-fashioned galvanised baths. Behind a small perimeter wall at the rear was a donkey boiler made from two large oil drums heated with a wood fire that provided endless hot water day and night. Two hoses led to the baths, one for hot and one for cold, so that once you had filled your tub you could sit in the warmth on the top of the cliff, your feet over the bath rim, discussing the excitements of the day and watching the sun go down and the stars come out. When you finished, the bath was upended and the water would sluice over the cliff and back into the river. Miraculously, no one ever followed the water over the edge, and only once did one of the tubs make the perilous descent.

At Mlawula we rediscovered the joys of communal bathing. A photo of our 'bathroom' once appeared in the American magazine *'Money'* as the most interesting bathroom in the world. Quite how it got there I have no idea. But there was one minor disadvantage with our bathroom. Every few days a troop of baboons would come and roost on the steep cliffs below to keep themselves safe from leopard, their arch enemy. During the night some of them would climb up and eat any soap that had been left lying around, with Palmolive being a particular favourite.

Leading off alongside the cliff, a path sloped down to a small platform built on a boss of rhyolite overlooking the bush and with a view down the Mlawula River. This was 'cocktail rock.' At the end of the day, coming back to camp around 5.30pm, we would grab a cold beer from the paraffin fridge and sit on cocktail rock to watch the sunset make its chromatic descent towards the western escarpment as its colour gradually transformed from yellow to gold, deepening to amber and changing to crimson before it slowly fell behind the crest of the far mountain ranges, leaving behind a cerise afterglow that lasted right through the next beer. As the light faded, in the grass next to us a pair of Swainson's francolins cackled their harsh partridge-like evening call, abrasively, as though in a fierce argument about their sleeping arrangements. Overhead into the darkness flew half a dozen Hadeda Ibis on their way to roost, noisily crying their raucous 'HA! Ha-a-a, HA! Ha-a-a,' which signified the last tattoo of the day. This flock always had one bird that seemed to squeak when he should have been 'HA-ing.' Ralph reckoned it hadn't been taught to speak properly.

"That bugger on the left's got a speech defect", he would say.

* * * *

If I had to single out one memory of 'cocktail rock', it would be of a group of volunteers who joined our excavations in the lowveld one July, including Doctor Dick from Eureka, California and Jeff the dentist from Fairfax, Virginia. How could we forget this duo? When they arrived at the camp, Jeff climbed out of the Land Rover and came over to introduce himself.

"Hi, I'm Jeff," he said, "care for a drink?"

With that he pulled open the right-hand side of his knee-length shooting jacket to reveal, in the places where the cartridges were meant to be, twenty miniature bottles of spirits – gin, vodka, Campari and assorted malt whiskeys. After a second or two, he opened the left side.

"Or perhaps you prefer a liqueur?"

Slotted on that side were miniatures of Tia Maria, Cognac, Calvados, Crème de Cassis and Cointreau – a whole bar-full. He fitted well into the 'cocktail rock' scene. He was a great enthusiast and as a dentist, bad teeth had been good to Jeff.

But it was Doctor Dick's night at 'cocktail rock' that will live forever in the annals of the Swaziland Archaeological Research Association. Dick's American patients had given him a send-off party and a 'survival kit' which comprised a flashlight, compass, a spare pair of socks, a packet of dried soup, fly swat, diarrhoea tablets and a couple of yachtsman's life-saver flares.

Supper had long finished and we were experimenting our way through Jeff's liqueur jacket when suddenly someone remembered that we had forgotten to commemorate July 4th the previous day. A toast was made, glasses were raised and the assembly launched into an inharmonious rendition of 'America the Beautiful.' Just as we reached the line 'From sea to shining sea,' Dick pulled the lanyard on one of the flares and it shot skywards in a shower of sparks before bursting with a thunderclap over the Mlawula River and the bush. The whole scene became suffused with an incandescent mauve light as the flare slowly descended. It made a very pretty scene. What we had forgotten, however, was that our Mlawula baboon troop were that night roosting on the cliffs below. The shouting, biting, screaming and howling from the cliff face lasted for half an hour, as the big males took to beating up anything within arm's length. You could hear the crash of baboons scattering in all directions.

It was some weeks before their nerves had steadied enough to roost below the bathroom again. I don't know what the baboon for 'purple' is, but they must have talked about it a lot in the intervening days.

* * * *

Baboons may be quite a long way down the primate ladder from modern humans, but if you watch baboons long enough you can somehow recognise all your friends. A big male can sit and observe you for minutes at a time, yawning and raising his eyebrows, just like someone you know. They travel as a troop of up to thirty or more, the females in the middle with the young riding on their backs, and the males around the edges of the pack. When they are on the move and while the alpha males are distracted by potential predators, the young males try to have their way with the females. The baboons were a constant source of interest, and during our bush walks the 'wahoo!' call of the big males echoing around the Lubombo valleys was very much one of the great sounds of Africa. Mlawula would be a great place to study primate evolution, and in this regard a very attractive young primatologist once approached me from the University of Illinois at Champagne Urbana. She was looking for a troop of baboons to study for her PhD. She had heard that I was connected to the National Trust Commission and involved with national parks, so she asked me if we had any baboons. She was a real beauty, with soft china-blue eyes and long blonde hair.

"Baboons?" I exclaimed. "Why, we speak of little else at Mlawula!"

I talked to her enthusiastically about our various troops on the Lubombo, trying to remember any extra details I could to keep the conversation in play. She confided that she was really passionate about her research and would be in touch. Sadly, she never was.

"Think of that?" I said to Ralph. "I mean, you have to believe me. She was really passionate. That's what she said. 'Really passionate'! I tell you, she was a real cracker."

"Yes, Price Williams, I am sure she was." He paused, then observed, "Mind you, you know her problem, don't you? She was heavily into baboons! Wouldn't have fitted in here at all. Pity, eh?"

Baboons are quite an evolutionary gap away from humans. For one thing, baboons have tails. The great apes, like the gorillas and chimpanzees, to which we are closely related, are like us all tailless. That troublesome bone the coccyx at the bottom of our bottom is the vestige of a tail, but it's a long time since we ever had one. Tail or no tail, people talk a load of coccyx about how far we have evolved away from the apes, but actually we are remarkably close. According to the evidence of DNA, our nearest relative is the chimpanzee, specifically the bonobo, *Pan paniscus*, the pygmy chimpanzee, which is alas now restricted to the forests of Zaire. These first cousins of ours are amazing animals. They all greet everyone else in the group with a 'good morning' shag. 'Top of the morning to you.' Bang! Bang! Bang! 'How are you today?' Bang! Bang! Bang! They shag everything in sight …all the time. They even play with themselves if there's no one else to have a go with. In a masterly example of understatement, our mammal book says, 'This species has a promiscuous mating system,' and, 'In the middle of the day they sleep in trees'. No wonder. They must all be knackered by that time.

And speaking of fornication, an awful lot of it must have gone on at the Mlawula camp, though mercifully I never saw it myself; I had my own caravan well away from the visitors because I was told I snored like a traction engine, especially after an evening on the 'red infuriator.' Now and again, when I was forced to make a call of nature on a moonless night, I would hear rhythmic rustlings that didn't exactly sound like the wild. The untamed African bush must have acted as its own aphrodisiac and over the several years we were at Mlawula we brokered at least four marriages, and God knows what other *liaisons dangereuses*. I put it down to all that fresh air and good food, myself, or maybe it was my evening lectures on the reproductive habits of the higher primates. But always, after the exertions of the night came the calm of the morning, and breakfast.

I have never been very communicative in the early mornings, and breakfast at Mlawula was no exception. If Toko was staying down at the camp she would bring me tea and scrambled eggs to my caravan just after first light. It meant I could collect my thoughts for the day, work out where everyone was to operate, and avoid having to answer the clichéd questions about how I had slept, or worse, listen to the banal answers of others who seemed hellbent on reliving their own ghastly nocturnal experiences.

Taking breakfast on my own had the added advantage of not having to watch others eat at that ungodly time of the day. In the dull early morning light everyone stood listlessly around the fire, with the last wisps of smoke rising from the dead ashes of last night's party. The only time I wish I had been there was when one of our new American recruits bounded up to the breakfast table, smeared his toast thickly with peanut butter and coated the ensemble with a liberal layer of chocolate spread, only to find that the chocolate spread was in fact 'Marmite', a peculiarly English salty yeast extract.

"Sheeit," he exclaimed, "what is that stuff? It's truly nauseous! It tastes like a combination of sea water and axle grease."

* * * *

Our surveys in the lowveld were beginning to fit together a picture of the past. The early Bantu-speaking farmer-colonists of the Iron Age, the black races of southern Africa, had not really settled in these low lying areas because of the risk of tsetse fly and ticks, both of which brought disease to their domesticated cattle, a problem wild animals and hence Stone Age peoples did not have. The hunter-gatherers had lived within the natural environment and had made no attempt to dominate it, as later immigrant agriculturalists were wont to do. I was especially interested in the microlithic cultures of the Late Stone Age and in our earlier surveys we had located many sites where these prehistoric groups had fashioned their tiny bladelets. These Late Stone Age people were the ancestors of the KhoeKhoen, the Hottentots, and more especially the San, the Bushmen, who have survived on the edge of extinction in the remote parts of the Kalahari of Botswana and Namibia to the present day.

The term 'Bushman' is a modernism. Even the name 'San' is foreign, a collective name used by early herders for people 'not like themselves' – 'Sonqua.' Indeed the hunter-gatherer San/Sonqua neither needed nor imagined a collective name for themselves as distinct from anyone else. The South African writer Laurens van der Post, self-appointed guru on the San, implied in his books that they were forced into the seemingly inhospitable areas of the central desert of Botswana where they currently survived the depredations of incoming black farmers from north of the Equator. The truth is even more tragic. The San/Sonqua were the indigenous peoples of the whole of southern Africa. They were once everywhere, many tribes of them, all with different languages, which in itself implies they had been living in their separate parts of the sub-continent, including the Kalahari, for a very long time, for tens of thousands if not hundreds of thousands of years. It was the coming of farming that spelt the death knell for these gentle people. They were systematically wiped out, first by the blacks and then by the whites, both of whom coveted their land to grow crops and run cattle. But the San had owned nothing. They lived in a spirit world, a world of imagery and trance. They had no concept of the rapacious imperative of tenure or title. They had no construct of possessing, no thought of keeping what they saw as a part of the world around them. To think of owning the land would have been as alien to them as attempting to own the sea or the sky. These things were not ownable. They were not theirs to buy or sell. But when the San refused to cooperate with the cattle-keeping newcomers, they were driven to the edge of extinction.

It was a human process which has been repeated throughout the globe, when the new order of farming clashed with the old order of hunting and gathering. For example in North America or Australia, U.S. committees and British politicians cared nothing for what they deemed to be peoples who were genetically inferior, indeed, barely human. They were to be aggressively exterminated like vermin, just like the San. And, as with the San, and with the native Americans and indigenous antipodeans, so died the shared intelligence and human observations of millennia, a profound insight into the great cosmic secret of nature, of an integrated creation. But these 'primitives' were not some inferior race; they were our very selves, imbued with the collective wisdom of the ages that has now, alas, been eradicated.

One ill-informed vision of hunter-gatherer peoples suggests that they lived a life that was short, hard and brutish, eking out meagre resources in a constant battle for survival. But this cannot have been so in Swaziland. Without the rapine exploitation of modern cultures, and surviving in mobile and comparatively small numbers, these Late Stone Age peoples would have had plenty from which to make a living. You only have to read the diaries of adventurers like Henry Francis Fynn who lived in Zululand at the time of Shaka Zulu, only two hundred years ago, to see that similar areas to Swaziland were renowned for their large herds of elephant and other game even that late. Before white hunters ravaged it, the Swaziland bushveld would have been the same, with elephant and all the supporting animals aplenty. Elephant knock down trees and open up the savannah for the smaller browsers and for the grazers. Game must once have teemed through these knobthorn savannahs. So much more must that have been the case before the coming of agriculture a thousand or more years earlier. The Late Stone Age way of life, though nomadic, without houses, villages or crops, was an ultra-specialised one. They had learnt every nuance in the bush. They knew every shadow, the call of every bird and habit of every reptile, and the migration routes and bush tracks of every animal, the times and seasons, the light and shade of life itself. It would have been a paradise for them. And if one area failed, they could always move somewhere else.

From what little knowledge we have of the surviving Bushmen, life for the Late Stone Age people must have been comparatively easy. When they were in small family groups they caught small game with snares made of cord crafted from the *sansevaria* plant, 'Mother-in-law's tongue.' When beaten, the leaves of this tough bushveld plant, which still grows today on the Lubombo, opens to reveal a mass of stringy fibres. By scraping them clean and twisting them together, a perfectly serviceable and strong twine can be made in minutes. Used with the branch of a springy bush, it can be triggered to close a noose around the leg of a small antelope or bustard.

Knowing the path each animal makes, knowing its peculiarities and its needs, they could read the bush like an open book. And to ensure their own future, they took only what they wanted. Their aim was never to kill more than they needed. Even their snares were placed in such a way that a bird could fly into a tree once caught, or the animal sit in the shade, undamaged and out of danger. If when they arrived at one of their traps the hunters had already caught their ration for the day, they would release the animal unharmed. If they were still in need of food, then before they killed it they would tell it a story, about the bush, about the divine presence that governed them all, and how they were grateful to their brother or sister for giving its life for them to live. They believed that all things were a oneness of the same life, the same world, the same universe. Their respect for the nature which surrounded around them was boundless.

The same respect was shown to larger game hunted during times when families came together and there were many mouths to feed. The men used small bows and short, lightweight arrows tipped with tiny slivers of shaped stone. It was not the brute force of the arrow that brought the animal to its knees. It was meant to be the skill of the hunter, his ability as a tracker, his stamina in the chase and his understanding of and empathy with the animal he was hunting. The San, like their Stone Age forebears, used plant poisons to tip their arrows.

They manufactured these in remarkable ways, knowing as they did about every tree that grew and every insect that crawled. One very powerful toxin they obtained was from the pupae of a beetle the size of a ladybird. These beetles feed on only one specific tree and no other, a *commiphora*, which incidentally is part of the myrrh family. It still grows in the rocky parts of the Lubombo today. The beetle lays its eggs on the leaves. The larvae hatch, eat the leaves and dropping to the ground burrow under the earth and form a cocoon. It is the tiny drop of fluid in the body of the pupa which is toxic, so toxic that an arrow tip smeared with this fluid has only to pierce the skin of the animal and within a matter of hours even the largest antelope, eland for example, will succumb. But there is a tree that produces a poison even more toxic. It is known locally as Bushman's poison and grows widely in the bushveld. The flowers are very sweet smelling, but if the wood is pounded and the sap boiled to a tar, an arrow tipped with this tar will cause death almost instantly. Even meat cooked over a fire made with the twigs is said to cause death in humans.

There are two questions everyone asks about the poison arrows. How is it that the hunters were not afraid of accidentally scratching themselves with the poisoned dart? Cleverly, they don't actually put the poison on the tip, but just behind it, so that the point has to be under the skin before the poison can reach the blood stream. And how is it that once the animal is dead they can eat its meat immediately without being poisoned themselves? In the case of the liver, they even eat it raw. The answer is these plant poisons are haemotoxins that only attack the bloodstream. The digestive tract is not affected. At least, that's the theory.

From the little mythology that remains of these vanishing people, the world around them was inextricably linked to their spiritual world, particularly to the animals that surrounded them. In their understanding about life and death, well-being and sickness, belief and ritual, the animal kingdom was a place where the material and the spirit world blended together. They juxtaposed animal and human behaviour. The actions of the eland or giraffe or kudu were minutely observed so that their behaviour was perceived to parallel human emotion and social relationships. The 'Bushman paintings' in the Komati Valley we had visited were an illustration of the same phenomenon, of the close proximity that these people felt to the natural world. They were integrated with it, not at odds with it.

But the bounty of the bush was not confined to animals. The trees and bushes provided a rich harvest of their own. One of the most important would have been the marula, a common bushveld tree which produces prodigious amounts of fleshy fruit not unlike apricots. The pulp contains four times more vitamin C than orange juice, and the stone is full of protein. More significantly, when the fruit is ripe it ferments naturally into a potent alcoholic drink, which I have no doubt that the San knew all about. Even today the local people make a wine from it – a sort of 'amber infuriator.' Recently it has even been distilled into a commercial liqueur, Amarula Cream. But the marula was only one provider among so many. Once the rains have fallen, the grewia bushes are laden with orange raisins. Other species provide sweet plums, russet berries and plump nuts. Wild melons ripen golden in the veld, and squashes trail across the grassy ground, below which bitter-sweet tubers grow in the soft earth. It is and was truly a Garden of Eden.

Even from the tiny portion left to us we can perceive that the knowledge these hunters of the Late Stone Age enjoyed about the world around them must have been encyclopaedic. But imagine how much has been lost. Think of the accumulated wisdom and learning of many, many millennia that vanished when the San were systematically eliminated. They knew the animal and plant kingdoms with an intimacy we can only dream about and which our scientific instruments can only begin to measure, never to appreciate. Studies of recent San hunter-gatherers imply that the people of the Late Stone Age spent less than two hours a day in food hunting and collecting. The rest of their time was free. What has happened in our world?

CHAPTER FOURTEEN

RHINOS, UP A TREE

Love is like a rhino; short-sighted, but always willing to find a way.
Anon

Predator levels being low at Mlawula, restricted to the odd leopard, it meant impala numbers began to rise alarmingly, especially once the reserve was fenced. Impala are both browsers and grazers and were so successful at Mlawula that they threatened the very existence of other grazers by over-eating the grasses. The only interim measure was to cull them. This is one of the most distasteful things we had to do, but the meat was sold off cheaply to local Swazi families who in this way felt the reserve had something to offer them.

The expedition would take four impala a week to feed our growing team. They were shot on a Friday, dressed overnight and taken early the next morning to the Long House. Sue and Toko would butcher and joint them and put them in the deep freeze. The cats always knew when this process was happening. From the garden they could hear the knives being sharpened and would hurl themselves at the kitchen window in an attempt to get in.

Impala meat, if cooked or frozen quickly enough, is like very lean beef. It has no fat and has no gamey flavour. Sue had organised a system for the lowveld victualling whereby we would take down a number of pre-cooked casseroles for each evening meal, frozen in blocks. She had scoured the cookery books for enough recipes for up to fourteen varieties of dishes. She invented 'Roman impala,' with red wine and onions; 'Hawaiian impala,' with nuts and pineapple; 'Turkish impala,' with red peppers and tomatoes; '*impala cassoulet*,' with chicken wings and speckled sugar beans; '*impala au vin*,' 'impala curry,' 'fricassee of impala,' 'impala hotpot' and so on. These rich and complicated recipes were not to everyone's taste. When I saw one of our younger volunteers scraping most of my wife's lovingly made '*impala Bolognese*' into the fire, I asked what the problem was.

"Don't you have any normal food, like hamburgers?"

American fast food in the bush? We bought an industrial food mincer and it was 'Build your own Mlawula burger' time, tender strips of prime impala, ground to perfection and mixed to our own recipe with exotic herbs (parsley) and spices (pepper and salt).

Ralph felt it was important to bring in predators to reduce the impala numbers; the weekly cull was no substitute for the real thing. A number of donated cheetahs were brought in from South Africa, where they had been reared wild within a reserve. We were assured that they would do well on the open flats where impala congregated. When they arrived they were put in a large *boma*, a fenced pen near the workshops at Mlawula, where they could acclimatise to the area. They were exquisite, with beautiful honey coats and black markings, especially the black 'tear drops' near the eyes. But happy they were not. If we went near the fence they would rush at it, bearing their teeth and hissing. When they were released they streaked away into the bush and we never saw them again, but one by one they took out every ostrich on the reserve. We found half-eaten carcasses for weeks afterwards.

"That's a bloody nuisance," Ralph commented. "These buggers seem to prefer poultry to game."

When they had dispatched the last ostrich, they disappeared for good, probably into Mozambique. Our impala problem remained, and the culling went on unabated.

※ ※ ※ ※

Mlawula reserve was next to Umbuluzi Estates, a large sugar plantation higher up the river, which owned the wild piece of land on the other side of the loop. It was a sizable acreage and had game of its own with its own warden Seamus, a particular friend and now neighbour of ours. Self-taught, he was a superb naturalist who liked nothing better than spending every waking hour, and some asleep, out in the bush. He knew every plant, every tree, every reptile and every bird call. If we had difficulty with something like that, we would ask Seamus.

Seamus had engaged a deputy warden, one Nkambule by name, who was a lowveld man of unhurried habits. They each had a vehicle and would talk to each other on the VHF. Nkambule would call up Seamus in a painfully slow rasping English.

"Mister Seamus from Umbuluzi Nature Reserve! Mister Seamus from Umbuluzi Nature Reserve! Mister Seamus from Umbuluzi Nature Reserve! This is Simeon Nkambule. This is Simeon Nkambule. Come in please!"

Since we were all on the same VHF channel, we soon got tired of hearing this laborious message, so Seamus took Nkambule aside.

"Look, Nkambule, why don't we each have a station number instead of a name? You can be Umbuluzi One and I'll be Umbuluzi Two, OK?"

That's how Seamus came to be known as Umbuluzi Two, as in, "Umbulzi Two needs to talk to you. He's just been on the VHF." And Umbuluzi Two he remains to this day.

Among his many achievements, Umbuluzi Two was an accomplished guitarist and would often come up to the camp in the evenings and play, which proved popular with the women; his ballads, usually country music laments, always seemed to set their hormones aflutter. He tended to rush into relationships with his eyes wide shut, hoping to find Miss Right. There was a sad tale he told of a short stay in U.K. where he had gone to advance his studies. In his first autumn term, he had taken up with a delightful lady called Pam and they had quickly become very close. Towards the end of term, Pam asked Umbuluzi Two where he was going for Christmas and he told her he was staying in college because he didn't have the money to go back to Swaziland. When she invited him to spend the holiday with her and her mother, he instantly accepted. They lived on a rather chi-chi housing estate in Watford. Pam's mother was a little dubious of Umbuluzi Two, this man from the African bush, but made him welcome. By the night after Boxing Day, he felt quite at home, to the extent that when Pam's mum asked him if he would like to make himself comfortable before dinner, he said that he would. Thinking he was back in Africa, he promptly opened the French windows, went out and pee'd on the lawn. Consternation! The following day he was back at his university flat, ejected from the house in Watford having lost the affections of the lovely Pam too.

"They gave me the bum's rush!" he lamented. "They made it abundantly clear that my sort of behaviour was totally unacceptable in Watford!"

* * * *

Quite early on working at Mlawula we found the Late Stone Age Lubombo microblade culture. We were not initially looking for it, but it turned up as a spin-off from some work Watty was doing. He had become interested in 'pans,' the shallow muddy pools that are a part of the southern African landscape. He had identified lots of them; in fact, you could say he had pots of pans! The pans, usually no more than a few centimetres deep, probably represented the lines of ancient and long abandoned water courses. For most of the year they were dry, but when the rains fell they filled up with a puddle of water anything in size from a small pond to a large pool. Naturally the water attracted game coming to drink while it lasted, and among the game would be rhino and warthog.

Both animals are in the habit of wallowing in the wet mud, which kills the ticks that irritate them so much. The ticks live on grass stems, and as animals walk past, they jump onto their backs and burrow into the skin to feed on blood, making the skin itch terribly. As the rhino rolls around in the mud, the sludge sticks to the skin suffocating the ticks. Once the mud dries, the rhino finds a tree and rubs off the clay with its dead ticks.

In this way, these animals have become geomorphological agents, instruments of soil movement. By removing the wet mud from the pan and depositing it near their favourite rubbing tree they are constantly taking away small but significant amounts of silt from the pan. The pan deepens and broadens, holding more rain the next rainy season, which attracts more rhinos and warthogs, which makes the pan even deeper and broader.

The pans were also a focus of animal activity in the past; herds of antelope and others came down to drink. Naturally this acted as a magnet for Late Stone Age people. They too knew where the pools were. More especially they knew which ones would be filled with fresh water because they had observed the 'rain animal' marching across the veld, which they believed brought the rain. The 'rain-bull' was another way in which the San expressed their own proximity to the animal world, in this case the mythological antelope that brought life-giving showers.

Looking across the bush on a hot summer's afternoon, the black moisture-laden clouds begin to darken the sky, illuminated as they drift slowly across the landscape by brilliant forks of lightening. The storm is coming. On the far horizon the dark clouds let down their moisture in fine silver-grey streaks, diagonal columns of rain, like the limbs of some great creature as it walks across the veld. The rain animal is stalking the land, its dark cumulus flanks and its 'legs' of wetness there for all to see. Distant thunder heralds its path as it plunges through the savannah. A pulsing wind shakes the trees before suddenly falling light and under a glowering sky the leaves hang lifeless from the branches and the grass gives up its susurrus. As the 'rain bull' advances its shadow blots out the sun as the thunder cracks overhead. Heavy drops fall from the leaden sky, splattering into the dust, the first footprints of the storm. The heavens open and in a cloud burst the rain animal leaves its drenching spoor over the empty sodden ground and moves on.

How many times had the San watched the rain animal? How many times had they traced its tracks? Hundreds? Thousands? Generation by generation, millennium by millennium, they knew the habits of the 'rain bull' as intimately as they knew the seasons of the impala, or the cycles of the acacias. Their life, their very existence, was conjoined in this essential metaphor, inextricably bound with an allegory they knew to be true. They had seen it and proved it, time and time again. With their acute senses and accumulated knowledge they could track the rain buck in the same way they could stalk the eland, because it was always so.

When the black races first came to southern Africa more than a thousand years ago, with their cereals and their cattle, they were ignorant of the passing seasons and they asked the indigenous San to make rain for them, which the San did. The San told the newcomers where the rain came from, and where it was going, not because they 'made' the rain, but because they were intimate with the landscape and how the rains behaved. Their profound wisdom, gleaned for so many generations of observation and experience, enabled them to predict rain. Matsebula, from the Trust Commission, told me once that 'Ezulwini,' the Valley of Heaven, didn't mean that at all. He reckoned the 'Valley of Heaven' was a contrivance of the local hotels to add romance to their image.

"Ezuluweni is the place of the rainmakers," he said. "When Sobhuza I, our present king's great, great grandfather, came to this place nearly two hundred years ago, he asked those people to make rain for him, that is, to tell him when and where the rain would fall. Rain is the life-blood of our earth."

* * * *

Back in the Late Stone Age, their knowledge honed to an incisive sharpness, the hunter gatherers knew when the pans would be fat with game, and they hunted there. We asked the game scouts at Mlawula where standing water stayed in the wet season and they revealed many a pan deep in the bush. At each one there were microblades and the scattered debris of their manufacture. Now we began to work out what they meant. The tiny bladelets were driven from a chunky core no bigger than the top of a human thumb. When the angles had been prepared, the core was bound with a thong and placed on a stone anvil. A small horn point was held over the platform of the miniature core and struck with a bone hammer. The percussive force of the blow caused a minute sliver of stone to detach from the core. The sliver was no more than twenty millimetres in length, parallel sided and thinned almost to translucence. The thong was unwound and the microblade released. This is what we were finding in the bushveld. The artefacts were so small, and the earth so sandy, that the only way to detect them was to get down on our hands and knees and peer hard, eyes right at ground level. It made for an extraordinary sight, grown men inching forward across the sand, back-side in the air and eyes glued to the surface, picking up needles of glass one by one and putting them in a labelled plastic bag.

We went to pan after pan, some more noticeable than others, some close by our camp and some way down in the bush. Each yielded the carnelian and agate microblades and microblade cores. We now knew where to find them, and how they were made, but when were they made? And what were they used for? And how could we learn more how these Stone Age peoples used the pans? The answers we hoped would come from excavations. We chose our first pan and laid out grid squares around one side of it. This was going to be hot and hard work, out in the open for hours on end, scratching the earth and measuring the finds. The first pan we chose to excavate we had already named 'Rhino Pan.' When the white rhinos were first introduced into Mlawula, a couple of young bulls were to be seen hanging around near the pan nearly every day, even though the pan was dry. We had more than a dozen rhinos now in Mlawula, but we didn't often see them close up.

The pan was quite small, only about ten metres across, and the centre was hard, caked mud, having been churned up by rhino and warthog then baked by the sun. It was surrounded by a thin scatter of small acacias, the Nile acacia, and another species known as the horned thorn with sharp spines, each as long as an index finger. We had just taken on a new team of volunteers and I had laid special importance on the role they were about to play in this world-class research, and how accurately things must be excavated and recorded. I showed them some of the microlithic tools, and made sure they examined the microblades through a hand lens.

"Take your time," I said encouragingly. "There is no hurry. There are no predators, so you'll be quite safe. Just set about it calmly and we hope to get good results."

They had a site supervisor, Donna, who had worked with me on other excavations who would guide them through the process. I wished them well and sped off.

Later I went back to see how they were getting on. I rounded the corner of the bush track and walked over to the pan. The sight would have done justice to the aftermath of a battle! The team sat about in a daze, bleeding from gashes to the legs and arms, neckerchiefs wound around their temples and clutching water bottles that had been emptied trying to clean their wounds.

"It was the rhino!" they chorused. "We were attacked by loads of them. They were coming at us from all directions, so it was every man for himself. We all shinned up the nearest trees and narrowly escaped with our lives!"

They had clearly had an intimate encounter not so much with the rhinos but with the *Acacia grandicornuta*, the 'horned thorn'.

"But these rhino don't charge," I countered. "They are white rhinos and are normally very docile."

"Well these ones charged! Or if not, then they certainly looked as if they were about to; they looked aggressive and hungry! It was terrifying!"

"How do you mean they looked hungry? They are grazers. They eat grass, not people."

"We didn't know that, did we?"

I pieced together what had happened. As they were sitting in the shade for lunch, a couple of rhinos had ambled over to see what was going on. They have poor eyesight, so they had come quite close. They had probably been standing there for five minutes or more before someone spotted them and ran screaming up the nearest tree. In the ensuing panic everyone else did the same. They stayed up the trees until the rhinos got bored and wandered off.

That evening, after extra rations of the *vinho collapsico*, Ralph explained that the rhinos, having not heard a vehicle, were relaxed and inquisitive. Suitably mollified the next day the team was back at work and during the two weeks they were there the rhinos came to visit every day, until they were adopted as mascots. There must be more pictures of 'Hey, Mum, this is me with our very own rhinos; see them behind the tree?' from the excavations of Rhino

Pan than any other in Africa. I was never sent any of the other 'This is me up the tree' pictures. Maybe they had been too scared to take any.

That same team had its fair share of intimate encounters with wild Africa. There was one rather bulbous woman who washed her 'smalls' at camp every evening and hung them on the line to dry overnight. One morning, much to her annoyance, they had all gone missing. We found them later in the bushes near the camp. They had been snatched off the line during the night by hyenas which had bitten enormous holes through each one with their huge fang teeth.

* * * *

"Well, Price Williams, you seem to have cracked the pan problem. So you've got a scraper industry, a microblade industry, and an MSA industry. How do those fit together?" asked Ralph.

"And what about my 'Christmas Boots,' the flakes I've shown you? Where do they fit in?" enquired Johnny, who was down for the night.

"They are all there," I stalled, "but they're all in the open, sometimes even all mixed together. What we need is a really good cave sequence that has the whole lot in, one by one."

"Well, we've got lots of caves in the Lubombo you know, Price Williams, so you'd better find one and get on with it!"

So we did.

CHAPTER FIFTEEN

THE PIPLESS LEMON SQUEEZY

The Original Lemon Wraps Inc. is one of the trusted names in restaurant and kitchen supplies for restaurant, kitchen, hotels, bars and other food service businesses.
Directory for Dallas, TX

I was walking with Watty along one of the remote spurs of the Lubombo crest in the exhilarating air of a cool winter morning. The black monkey thorns grew stunted and tightly-branched along the unyielding rocky edge of the scarp. A scaly-throated honey guide purred its ascending rasp in the bushes off to our left, and black-crowned tchagras whistled their haunting call in the deep ravines. We were searching for caves in the canyons that cut into the scarp face. We had already examined a few cracks and cavities among the steep rocky slopes, only to find them too small, or their sloping floors filled only with bat guano.

"You know how these caves are formed?" said Watty. "When the rhyolite is extruded as very hot rock dust, it's not a single incident. It's cyclic - extrusion, then calm, then extrusion, then calm, twenty million years, switching from one to the other. Today the caves develop where the sediments which formed during those calm phases are exposed and eroded."

It turns out that in the calm periods, even the hostile and barren volcanic surface started to weather. With rains driving in from the newly formed Indian Ocean, pools began to form and plants started to grow in the mire around the edges. Sometimes the ash-fall events were a million or more years apart, plenty of time to build up some muddy sediments and preserve fossil plant remains. Watty had already found some with frond-like leaves, from a plant called *otozamites*.

"Dinosaurs fed on these leaves, right here." He pointed to the black leafy casts in the stone. "They are plants directly related to the cycads that grow here today. These little pools form on top of the last lot of hardened ash. When the next eruption takes place, the next ash-fall travels over the ground at such a speed it traps the little pools and their sediments under the ash, with all the fossils in, and bakes them hard. But the sediments are loaded with salts, and when that little hard puddle of salt-rich sediments is finally exposed by erosion, the salts come to the surface and the sediment crumbles away like dust. Result: caves."

The process of rhyolite formation was mind-boggling - the flaming clouds of gas and ashy dust, glassy lavas, even jets of red-hot liquid silica squirted high into the atmosphere which then fell back as glass rain. We had seen what looked like water droplets on the ground that morning, only to find they were made of glass that had solidified as it fell to earth. And all this had taken place 180 million years ago, when dinosaurs roamed Africa.

As we saw and rejected each cave, we finally came to one that was perfect for our needs. It wasn't very large, about ten metres across the mouth, and about four metres deep. Someone had poked a hole in the floor some years before and found very little. But that had been a weekend affair. Maybe they were looking for the missing gold Kruger Rands, a local sport in the Lubombo. It was rumoured that when Paul Kruger had gone into exile at the end of the Boer War, he had hidden millions in gold coins somewhere in these mountains on his way to the coast and to banishment in Holland. A closer look revealed no gold coins, but it did show plenty of tiny pieces of chalcedony microblades and even Christmas boot flakes eroding out of the front of the deposit. Properly excavated, this might give us the Lubombo sequence we were looking for, as event after event had built up into a thick layer of cave debris.

The cave faced north, across the narrow valley of the Siphiso stream, one thousand feet below the Lubombo heights. The low-angled winter sun streamed in across the mouth. We would have to do something about that. The bright sunlight and strong shade would obscure the delicate differences in the colours and the textures so vital in identifying the numerous lens-like layers that made up the thousands of years of ephemeral use of the site. Also the deposit was very ashy, the result of endless fires that had been lit by Stone Age people to cook their meat and to keep warm. The ash meant that we would be best advised to stop people walking on the deposit to prevent it becoming mixed together. A new group of volunteers was due to arrive that weekend anyway; they would be the first to excavate at Siphiso.

* * * *

We entertained many extraordinary characters passing through the Long House. Maybe the whole notion of volunteering to travel to the African bush to dig up prehistoric artefacts attracted more than its fair share of cranks and eccentrics, but they were all fascinating in their different ways.

Take Barry H. Lefevre, the 'Pipless Lemon Squeezy' supremo, for example. I first encountered him by letter. I had the habit of writing ahead to all the volunteers who hoped to join us in Swaziland. You know the kind of thing, 'If there is anything I can do to help before you travel…' Back from Barry came a letter.

'Dave, Thanks for your letter. Really looking forward to meeting with you. Can you tell me if I need to bring … Barry.'

But the striking thing about Barry's correspondence was not the questions he asked but the heading on his writing paper. Across the top, in large canary yellow type, was the following;

> THE PIPLESS LEMON SQUEEZY
> A Barry H. Lefevre creation.
> Barry H. Lefevre, Inventor, Designer and Sole Distributor.
> The Label for your Table!

The title was followed by an address in San Diego, California and a logo showing an animated lemon with a toothy grin on its jaundiced face.

I developed a keen interest in meeting this citric colossus. When he arrived at the Long House he turned out to be a small, balding, rather cherubic man in his sixties, much given to expansive gestures with his short arms and his finely manicured fingers. I soon discovered that the magnanimous hand waving and dismissive cries of 'Po! po! po!' were the result of him being of French Canadian extraction. He exhibited a certain Gallic flamboyance and still spoke with a hinted modulation of *quebecoise*. His wife, Boopsie, was from the East End of London, with an accent to match. I had my first chance to catch Barry in person at dinner that evening.

"Tell me, Barry, I hope you don't mind me being inquisitive, but may I ask you, what exactly is the 'Pipless Lemon Squeezy'?

"*Mon Dieu*, Dave, you have not heard of the Pipless Lemon Squeezy? And you in the catering trade and all!"

We were twenty five people for dinner that night.

"You get a piece of calico, say yellow for preference, and you cut it in a circle, about yeah big." He described the shape of a side plate with his two hands,

162

index fingers and thumbs outstretched. "Then you take a lemon and you cut it in half around the equator and upend it on the calico. You draw up the edges and tie them at the tit on the top with a ribbon, you know, preferably yellow. *Eh voila!* The Pipless Lemon Squeezy!"

"And what do you do next, Barry?"

"Why Dave, you squeeze it on your fish. And with your lemon enrobed in our Pipless Lemon Squeezy, you get no unsightly pips falling on your dish and you don't get lemon juice squirting in your eye either. See, Dave, there was this chef in San Diego who posed me a problem and that was my answer. I used to cut each 'squeezy' by hand with pinking shears in our garage, and Boopsie, God bless her, used to cut the ribbons into lengths, didn't you Boops? She's a bit deaf now, but great on ribbons she was. *Mon Dieu!* No one better! Eh, Boops?" He smiled at her indulgently.

"And you made money doing that?" I asked incredulously.

"Oh sure, Dave. But hard work you know. Up all hours in the garage, snipping away. Especially after the Hyatt Hotels took the product, and then the Hilton chain. Sure you haven't seen them, Dave? They are in every restaurant in America. Well, it all got too much for me, and Boopsie, she got fed up with the whole thing. When you gotta snip five thousand pieces of ribbon a day, it gets to you. And her poor fingers. So I sold it last winter to my brother-in-law. Should have got more for it really, but we were happy with the two million he paid, weren't we, Boops?"

And then there was Fred, a tall gangling character in his eighties, who looked and dressed as though he was straight out of a 'history of American vaudeville' tableau.

He had a little banjo and without prompting would accompany himself, singing medleys he had written over the years, a sample of which might include: 'Oh golly, oh gee-ee-ee, you're the only girl for me-ee-ee!' Just in case some African impresario wanted to snap him up, even at this late stage, he had brought with him a case full of rolled up and dog-eared scripts and renditions which had yet to be performed. Alas he never did feature on anyone's audition list, but it didn't seem to dent his indomitable spirit. He just kept on strumming.

Picture him on the first morning at breakfast at the Long House. He came downstairs dressed for the bush. He had abandoned his theatrical gear and was wearing a capacious pair of khaki shorts that contrasted greatly with his bandy legs and extremely knobbly knees. He laid out on a table thirty one containers, each one with a set of brightly coloured pills inside, then emptied one lurid pill from each and devoured them one by one. Then he took a bowl of cereal and from his pocket withdrew a small, plain silver box and with enormous concentration slowly rotated the box over the bowl, first one way then the other.

"Say, Fred, what the hell are you doing there?" asked one of the younger volunteers.

"Oh, I'm de-ionising my cornflakes, Randy. Get too many negative ions and it plays havoc with your vocal chords. I need to keep my larynx in trim. Cornflakes are specially damaging, you know, if they're just left like that. Should be a warning on the packet!"

"So what's in the box, Fred?" I asked.

"Ah, well, that's kinda secret. If we tell everyone, no-one's gonna buy one are they? And we wanna sell! See, there's a crying need for these, so my partner says. After I pestered him for a long while he sold me twenty percent of the company, and free boxes for life! It's a really good deal. See, here's what this box can do."

He handed me a scruffy piece of printed paper. It read.

'The Magnetic Way to a Better Life! The magnetic qualities of the carefully selected herbs in this box are guaranteed to increase vitality, de-ionise harmful zeta rays, prevent foot odour, discourage venomous serpents and other reptiles, improve short term memory and reduce the speed and strength of hurricanes. Also believed to be efficacious in the treatment of verrucas. DO NOT OPEN.'

"See, it really works. Look at me! I'm over 80. I've been de-ionised for years!"

That's what we all thought.

* * * *

We drove down to Mlawula, a convoy of Land Rovers. We had five by now, and had sprayed them all in the same livery, brown and cream, with the SARA logo on the doors, showing the sun setting behind a knobthorn from cocktail rock, with the mountains along the bottom. The vehicles needed constant attention because of the hammering they were getting in the Lubombo, and we went through tyres like a kid with balloons. I had even brought out a student from London who also happened to have experience in automotive engineering. He was short, but wiry and tough. He hailed from Edinburgh and became known universally as 'the Wee McScott.'

The Wee McScott was particularly taken with one of the volunteers, Martha, part Cherokee Indian and, by her muscle tone, part Amazon, who liked to watch him fixing up the Land Rovers. After he had worked all evening on one particularly quarrelsome gearbox I asked him:

"How did you get on?"

"Usual," he called, from inside the engine, "left leg over first, you know!"

The Wee McScott clearly serviced more than just Land Rovers.

At the cave we erected a black shade cloth across the mouth, held on hawsers stretched outside from the overhang to the apron, with a door at the one end. Inside, we had arranged scaffolding over the deposit so that you could lie on your stomach on foam mattresses and excavate square metre by square metre without touching the archaeological material except within each square. A wire grid was suspended from the ceiling of the cave laid out in metre squares lit by fluorescent lights powered by a small generator running outside the cave so that it couldn't be heard. The idea was that we would excavate four one metre squares at any time, two people to each square. With finds recorders and a site supervisor, the complement was usually about twelve.

The aim behind all forensic excavations is to isolate each layer by it colour and texture, in the reverse order in which they had been deposited. Each layer in its way is a tiny time capsule. Everything within it ought to be contemporaneous with everything else in that layer, and relate to the life and times of the people who lived there for that brief moment years in the past. That is why the control of the layers is so critical. In a general sense, layers that are lower are earlier. Layers that are higher are later. The hope is that as each layer is separately exposed, with its tools, bones, beads, charcoal and whatever, that group of material will tell us something about that instant in time. By recording the exact position of everything within each layer and within each square, all three dimensionally, it would theoretically be possible to replace everything exactly where it came from, or rather, we would have a permanent record of the relationship of all the artefacts in all the layers.

This is a slow, meticulous task. At Siphiso, it was made more so because the deposit was so dusty, and we had to expose each artefact and each specimen of bone not with trowels, as one might in an outdoor site, but with airbrushes, gently puffing the dust across the square to reveal the debris, recording each tiny piece with horizontal and vertical co-ordinates, then bagging each item in a plastic Ziploc. All the details were recorded on one sheet per layer per square. To test the layer colour we used soil colour charts and entered the code on the sheet. We would later also analyse the particle sizes of individual grains of sand, silt and clay in each layer as an extra control on our layering.

They were quiet contemplative days for the diggers, or more correctly I suppose, the puffers, as they lay on the boards under the soft lights, puffing the dust away and reciting the coordinates as the remains were removed piece by piece. Most of the time the work was routine, bagging each sliver of bone, each artefact, each piece of charcoal. Now and again something really significant would turn up, like a diagnostic stone tool or an especially important bone or tooth. We would stop the excavation and troop outside to sit on the rock under the dry, leafless marula tree that was our day-time home. I would then describe the significance of the find in the context of the layer we were excavating – 'this is a side scraper, this is an end scraper, this is the toe bone of a klipspringer' and so on – so that everyone felt motivated to continue puffing until the next coffee break. There were toilet breaks too, when you could take a spade and loo paper and walk around the back into the woodland. The loo paper was kept in its own 'finds bag' labelled 'end scrapers!'

We had some difficulty finding the right equipment for blowing the dust away. Anything automatic cost a fortune and would have become inoperable with the grit. Then someone suggested enema tubes, which turned out to be perfect for the job, a long narrow plastic tube with a large rubber ball on the end that could be squeezed as gently as required. With some difficulty we had found eight of them in various pharmacies in Mbabane. The problem was that with constant use the rubber would split and we needed new supplies. I went back to one of the pharmacies.

"Excuse me, do you have any enema tubes?" I enquired.

"Yes we do have. What size do you want?" The young Swazi girl behind the counter was all smiles.

"The largest you've got, yes, that one with the large ball. Do you have eight of them?"

"What? How many did you say? Eight?" Her smile clouded over.

"Yes," I persisted, "eight. You see, because we use them all the time they do wear out very quickly, especially when it gets really stony."

The Swazi assistant fled into the back of the shop and pulled the pharmacist out to the counter.

"What seems to be the trouble here?" he asked severely. "What are you doing to Africa?" He motioned towards his assistant. "She tells me you have a strange request!" I puzzled for a moment over these seemingly unrelated statements, but Africa turned out to be the name of his Swazi assistant, Africa Hlope, I read on her badge.

"I simply need eight enema tubes. We are wearing them out at the rate of one a day at the moment – obviously that is on weekdays. We take a break over the weekend."

He showed a look of distain and alarm by turns. He must have been imagining scenes of communal colonic irrigation by a coven of European perverts.

"We only blow air, of course, not water," I said, by way of clarification, which made matters worse. Then a light seemed to go on.

"Aren't you that archaeologist fellow that's working down the lowveld? So that's what this is all about! I'm with the Swaziland Ornithological Society, SOS for short. We always try to help. Glad to meet you. Enema tubes? Certainly, I'll put in a permanent weekly order for you. The rep. will be here tomorrow. I'll tell him."

He leaned over and spoke confidentially behind his hand.

"The rep. won't be surprised. He thinks we're all full of shit anyway."

* * * *

Back at Mlawula the excavation of Siphiso had begun. Softly the dark brown gritty dust was being blown from the first layer, the top of the deposit. In Britain this would be the one with the dud batteries, old sweet papers and discarded contraceptives. In the remote bush it contained the remains of fires, a few shards of pottery, a chewing gum wrapper and whittled wooden pegs that were probably to be used for modern snares. The broken pottery was a coarse ware with indentations on it, probably from a small beer pot. It had been fired in a bonfire. The surface colour was blotched from brick red to almost black, typical of the Late Iron Age, which could date from the last five hundred years ago right up to the present day. The date was confirmed by the discovery of a small green glass bead, like the ones that were traded by Europeans in the 18th century. In other words, it was from the recent use of the cave by local people. For the first few days, that was all we were to discover.

At the camp things were running well now, as everyone knew how everything operated. Or that's what I fondly believed. Until Thursday morning when Toko came to my caravan at dawn with tea and toast, but no eggs.

"Toko, are the eggs coming?" I asked.

"*Kuti emacandza, Bahbe*! There are no eggs left."

"Toko, there must be. Sue wouldn't make a mistake like that. It's only Thursday."

"I know *Bahbe. Buka wena! Uyaphelile*! You look! They are finished!"

Sue had the camp provisioning down to the finest of arts. So many people for six nights, that's six breakfasts, six packed lunches, six dinners. One or two eggs for breakfast, maybe a hard boiled for lunch, times six days, times twenty four people, that's two dozen dozen. She had done it so many times before and we always had enough. Our army not only marched on its stomach, it also was beginning to excavate on its stomach. I was still arguing with Toko when Boopsie overheard and said.

"Dave, you know I was off sick yesterday. Well, after you had left for the site Jabu was cooking an omelette."

Jabu was an extra maid we had taken on for a couple of weeks because our numbers were high.

"She was cooking this omelette, and I thought, 'My! That's a big one.' She was cooking for her, and for Phillipe and this whole load of your game scouts who turned up, and they're all eating and eating. It was quite a party. She must have used up twenty eggs or more. It's not healthy, that, is it?"

Jabu admitted she had used them. There were so many eggs in the box, she had never seen so many, ever. We must have brought down too many and she was afraid they would go off. I had told her to help herself to anything she wanted, so she had.

She looked so innocent, and indeed she was. This was such a typically Swazi thing to do. In a time of plenty, everything is shared with everyone else. Seeing so many eggs, she invited all and sundry to join her in her late breakfast. We all had a good laugh about it afterwards, but at the time we were eggless. At least it might save on the enema tubes.

CHAPTER SIXTEEN

GENERAL WEAKNESS

Our greatest weakness lies in giving up. The most certain way to succeed is always to try just one more time.
Thomas A. Edison

In the early light before the sun had risen above the Lubombo crest those winter mornings began cold. We were palpably warmer at the top of the little Lubombo ridge, but driving down the rocky track into the valley the air became progressively frigid, with a chill vapour settling in the hollows. It added its own prehistoric atmosphere, seeing the stark, bare branches of the thorn trees looming out of the murk, with herds of impala half-hidden in drifting mist. The Land Rover jolted on along the track, past the gardenias and confetti bushes, the branches squeaking along the sides as we went. Reaching a clearing at the head of the valley, we got out and arranging our backpacks and equipment, began the walk up the dry gully to the cave. It was usually a silent affair, that fifteen minute morning walk.

Each of us had our own thoughts of Africa. The gravel of the streambed crunched underfoot as we brushed past thickly-fronded wild date palms, groves of milkwoods and the ever-snagging river climbing acacias that criss-crossed our path. Monkeys sat in motionless rows atop the tall branches of the ashen-barked fig trees, all facing the sun to catch its first warmth, keeping a look out for the martial eagles that wheeled over the valley ready to swoop down and pluck one of them skywards. The last part of the walk was up and over a dry, rocky waterfall, behind which were pools full of clear water. This must have been the source of water for the Stone Age people. The last uphill climb from the pools brought us to our marula tree and the cave mouth.

We all stood in a line for a few minutes, facing the rays of the mounting sun to absorb its warmth, just like the monkeys. The generator was switched on and we would disappear behind the shade cloth to begin the day's excavation. We had an accomplished site supervisor for Siphiso, an elegant, quietly-spoken American PhD student, Larry, who had endless patience and a fine grasp of the Stone Age. Toko, who like many Nguni speaking people, could not distinguish between 'l's and 'r's, always called him Rally. Larry once invited Toko to come and see what we were doing at Siphiso. She had never been to an excavation, and although she had seen all the stone tools lying about the Long House, hadn't really thought too much about where they came from. She walked up to the cave and came inside where everyone was lying prostrate and puffing among the world of the distant past.

"*Hawu! Bahbe, Hawu!*" she exclaimed shaking her head and shrieking with laughter. "*Hawu! Bahbe*! What are you doing here? You are all playing in the dust like children! Just like my little boy!"

Larry took her to the side of the cave and patiently showed her what we had found and where it had come from. Toko quietened and screwed up her face to try to understand; she had great respect for Larry. But the concept of distant millennia was foreign to her. Life was today, right now. Ten thousand years ago for Toko never existed. From then on she tolerated us playing in the dust, and washed my shirts without complaint, but whenever I mentioned the cave she would smile knowingly and say nothing.

* * * *

After six days the first Siphiso team had eaten their fill of the Layer One dust, so we took them on a journey of exploration down the Mbuluzi gorge. The contrast was striking. If the Nkumbane valley and the bush were quintessential 'game viewing' Africa, with distant vistas, savannah grassland, flat-topped umbrella thorns, giraffes, rhinos and herds of zebra and wildebeest, the gorge by contrast was tropical Africa, '*Sanders of the River*' style steamy rain forest with lianas, crocodiles and red simango monkeys. The precipitous sides were thickly forested and the Mbuluzi flowed sluggishly through steep, shadowed spurs and dark cliffs, overhung with Lubombo cycads, euphorbias and dripping mosses. As we entered the gorge there were a few wooden ashtray trees, the ones with the lurid red and black seeds. Pressing on along a rough track that was annually washed away by the torrential summer floods we came to the narrowest opening of the gorge. The railway from the Ngwenya mine joins the gorge at this point, built on a high embankment above the river. It had been constructed to take the iron ore from the highlands of Swaziland down to Mozambique and the port at Maputo from where it was shipped to Japan. Amazingly they were still using steam engines on this line.

At our camp at Mlawula it was a sensation tinged with nostalgia to lie awake on a star-spangled, still night listening to the bush sounds, the jackals, the hyenas and the nightjars, then to hear far away on the night air the echoes of a locomotive in the gorge, the steam hissing from the cylinders, the chuffing from the smoke stack, then the dull rattle of the truck wheels groaning on the rails. There was something otherworldly about it, reminiscent of Africa in the nineteen twenties, like the Kenya of Karen Blixen, Elspeth Huxley or the Happy Valley set.

The trail down the gorge was always difficult to navigate. The washed away sections had been only roughly filled in with boulders, and there were always times when everyone had to get out of the Land Rovers and hurl stones into the side streams so that we could drive across. In the hidden curves of the gorge there was a rich collection of rare trees and plants; the wooden microphone tree, toad trees and the wooden bananas. The fruit of the wooden banana looks exactly like its name-sake, a banana made of wood, with the wooden peel curling backwards. They are imposing trees that grow up to one hundred feet high.

Deeper down the gorge, Nile crocodiles lay menacingly on the boulders and stretched out on the sand banks. Some were massive. They spied us toothily with glaucous eyes before they slid silently into the depths of the midstream pools. A little further on we came to one of the most alien phenomena of all along these Lubombo gorges, the Lubombo ironwood forest. At first the ironwoods appeared in ones and twos before crowding into a dense mass of black trunks canopied with dark evergreen leaves. As we drove on, both sides of the gorge became thick with them.

We made a special study of these strange trees, because in all of Africa it is rare to see what was, in effect, a natural forest of a single species. Most parts of the savannah have a dozen or more species competing for space, but the ironwoods stood alone in the gorge. It seems they secrete a toxin in their leaves that poisons the ground; local families say that wild honey made from the ironwood flowers sends people mad. No termite will attack the dead wood and very few other plants chose to live among them. But one that did was the Mbuluzi cycad, *Encephalartos umbuluziensis*, which only grows in this one ravine among the ironwoods, and nowhere else in the world, another relic from the days of the dinosaurs. This cycad grows at ground level with only the leaf crown showing with recumbent fronds of thick, palm-like, spiky leaves, in the centre of which are the long yellowish-green cylindrical cones.

I once took a group of visitors into the ironwood forest to see the cycads. They were a garrulous bunch and disinclined to listen when I spoke to them. I had just announced that we would soon have lunch, but first, I said, "Let's go and visit the cycads." We drove into the forest, got out of the Land Rover and walked among the trees.

"This plant, the Mbuluzi cycad, is one of the most remarkable plants in Swaziland," I began.

Stan, a time-serving attorney from Philadelphia, turned to the woman next to him and whispered, "Say, Nancy, is this your sandwich?" He held out a polythene packet between his thumb and forefinger.

I continued in a louder voice. "We find traces of cycads as far back as the Carboniferous period, when the coal measures were being laid down."

"Oh, it isn't your sandwich? Well, I didn't know that."

"The botanical interest in cycads is that they are cone-bearing plants."

"Betty! Say, Betty, is this your sandwich? What? I don't know what kind of sandwich it is. What d'ya know? It's a chicken sandwich. You wanted chicken, is that right?"

"This particular cycad grows only here in the Mbuluzi gorge and nowhere else in the world!"

"Yeah, its chicken…with, uh, mayo!"

By now the whole group had eyes and ears only for Stan and the polythene bag.

"The dinosaurs fed on these sandwiches!" I mumbled weakly.

"Hell, David, that's so interesting. What did you say that plant was called again? Cyanide, David ? Is that it? Cyanide?"

"Something like that Stan!"

We ground up one of the Lubombo tracks that led to the summit and worked our way over to the furthest fence line of the Mlawula reserve, the border with Mozambique. There is always some special excitement about borders. These arbitrary lines seldom have any topographic logic, but there is a *frissance* about thinking that here is Swaziland and just there is what was once Portuguese East Africa. There used to be a notice at the border post on the main road leading down to Lourenço Marques which read, 'You are now entering Portugal,' right there in the remote African bush.

Sue remembered going down to L.M., as it used to be known, for weekends when she was a teenager to eat the biggest prawns in the world, washed down with *vinho verde* imported from Lisbon. We could see L.M. now looking through the game fence, and could make out in the distant haze the apartment blocks along the palm-lined *avenidas* only forty miles away. Wouldn't that be a great thing to do, we thought, to enjoy a sybaritic weekend by the sea?

But it was not to be. Mozambique was gripped in a cruel civil war between rival African factions. There was no sign of the war where we were, nor did it affect us at all, except that now and again when the wind was from the Indian Ocean we could hear the distant boom of artillery. It would be many years in the future before I made my first foray down to the newly-named capital, Maputo, to enjoy the legendary foot-long prawns grilled in a fiery red-pepper sauce and the unique aroma and *petillance* of the tart wines of the valley of the Douro, and from the coast I could look back at the Lubombo to the place where we were standing that winter's day.

※ ※ ※ ※

"David, what do they mean by 'general weakness'?"

A new group of volunteers was driving down to the bush. We had just enjoyed a spectacular send-off for the last team. Someone had bought a couple of bottles of white rum in the duty free and forgotten to take them down to Mlawula, and as they didn't want to carry them all the way back to the U.S. we were invited to a piña colada party at the Long House. Larry had stopped at a bottle store in Manzini and bought a bottle of Coco Rico coconut liqueur, then stocked up at Mahlanya market with a sack of Malkerns pineapples. The stage was set for a quiet evening sipping a new cocktail, which turned out to be a wild evening of toasts celebrating Africa. Larry mixed the rum and coconut liqueur together with crushed ice, scooped out the pineapples, and served everyone with a monster coupe of the most captivatingly fearsome drink I have ever tasted, with palm fronds and other tropical bric-a-brac sticking out of the top.

So when someone asked about 'general weakness' the next day, I wondered whether they were referring to my own well-being, my head at the time feeling somewhat like the coconut on the front of the liqueur bottle. It turned out that we had just passed a notice on the side of the road, something to the effect that one Dr Zwane, a well-known *inyanga* and traditional healer, could cure warts, wind and 'general weakness.'

'General weakness' is the African equivalent of 'brewer's droop,' the inability to get-it-up, or if you could, to 'keep-it-up.' Among the Swazi, this can be a crippling social problem. Certain Swazi elders, as a perk for their participation in running the nation, were allowed to take a young Swazi girl as a nightly comfort. However, the practice fell somewhat flat if the aged warrior suffered from 'general weakness.' The botanical equivalent of Viagra was then administered by the *inyanga*, though quite how has not been revealed.

On one occasion I took a Swazi assistant named Khumalo down into the bush to show him some archaeology. He loved to flirt with the ladies, clad as always in his colourful traditional dress, a series of items not unlike a Scottish clan elder – kilt, sporran, shoulder wrap and so on. He spoke little English except a stentorian 'yeeess' and 'noooo,' not that it limited his sexual activities. I was woken one starry night to the rhythmic rocking of the springs of the nearby Land Rover. It was Khumalo indulging in what was known locally as *ukuhlobonga*, the 'delights of the road,' a dalliance with a passing female. The only difference was that in this case there were no fewer than three of them on the road simultaneously, all newly-arrived volunteers in the expedition team. They were women of a certain age and each night they drew lots to decide who should enter the Land Rover first. It was apparent that Khumalo did not need the ministrations of Dr Zwane as he clearly did not suffer from 'general weakness.'

Incidentally, *ukuhlobonga* might sound a great idea, but it can be a double-edged sword; as there are delights, so there are also penalties. If in the course of *ukuhlobonga* the warrior, say, Khumalo, happens to get a girl pregnant, he can be arraigned before a traditional court by the parents of the girl and fined heavily and made to pay in Swazi currency - cows. Khumalo admitted that this had happened to him on no fewer than three occasions. Each time he had been fined three cows. As he only had six to start with, he was in bovine overdraft and was unable to raise the necessary ten cows that would be required to pay *lilobolo*, the bride price, should he wish to marry. So, caught short of the ready heifers, he was forced to continue with *ukuhlobonga* for the time being. On this occasion the trio of ladies obliged.

* * * *

At Siphiso we were about to tackle stratum two. This was quite different from stratum one, marked by a distinct change of colour and texture. It was a thin layer of white dust, no more than two centimetres thick, right the way across the cave; it acted as a persistent marker which separated the modern from the ancient, almost like an undisturbed membrane over the Stone Age deposits. Larry was puzzled by the fineness and whiteness of this dust. We thought at first it might be the accumulated scat from hyenas that had used the cave as a den. In their dens they crunch up so much bone and pass it in the droppings that it can build up a substantial layer of calcium. Certainly the chemical analysis of the layer showed it was rich in calcium. But there were no little slivers of bone that would also have been part of the droppings, nor were there any larger pieces of feet or legs or ribs that normally would have been dragged into the cave. I suggested it might be the ash from the constant burning of calcium-rich wood, like the leadwood. But it didn't look like an ash. It was not soft enough. Finally, Larry reckoned that it was sheep dung that had weathered down to a calcium-rich tilth. There were parallels for this in other parts of southern Africa. When the Iron Age 'Bantu speaking' farmers ousted the San from their traditional hunting grounds, they brought with them fat tailed sheep which originally had come from the Near East. They penned them at night in rock shelters like Siphiso to keep them from being snatched by leopards. With perhaps a few decades of such use, and the overnight dung being trodden flat, then weathered, this was the result.

"Do you mean to say," one humourist suggested at one of the morning briefings when I was describing progress so far, "that I have come all the way from Detroit, Michigan, home of the great U.S. automotive industry, to shovel three hundred year old sheep shit? Wow! That's really great!"

Date-wise, he wasn't far off, archaeologically speaking that is. The radiocarbon dates for Stratum Two indicated that it was formed about three hundred and fifty years ago. These sheep had been penned up at Siphiso around about the time the Pilgrim Fathers landed in Massachusetts and founded Boston.

During these dusty days the evening bath on the cliff was much appreciated; by five in the evening everyone was covered in grime. We would wend our homeward way down the dry riverbed to the Land Rover, then drive back down the valley to the camp. There was always a scramble the last few steps of the walk to get the right place on the vehicle for the ride home. The coveted position was to sit outside. Two people would sit on the roof, feet dangling in front of the windscreen, and one person would sit on each of the two front wings, hanging on to the mirrors, and the rest would be inside. Slowly we would lumber down the track and the two on the roof would have to dodge the acacia branches as we went. But they got the best view of the game, being that much higher, and would sing out if they saw something interesting, like a male kudu or a couple of browsing giraffe. The two on the front had the advantage of feeling they were closest to darkest Africa. The only drawback came if we rounded a steep curve and a rhino or two was standing right in the middle of the track. It was at times like that, before what little brakes we had begun to stop the vehicle in a series of shudders, that the front riders felt that Africa was coming a little too close for comfort.

As we came up the last part of the incline, we would put on a burst of speed and skid into the camp; then it was an undignified scramble for the baths. Whoever was not first in the bathroom would migrate to cocktail rock with a sundowner and plate of peanuts for the daily sunset ritual. While there was still light, in the Mlawula valley below we often saw water-buck in the reeds and lines of zebra nodding their way back into the bush after their last drink of the day. The most comical was to watch the baboons that wanted to cross the river to get to their roost under the bathroom. They obviously hated getting their feet wet and would spend some minutes scampering up and down the sand banks trying to guess the best stepping stones to cross. Most managed easily, but now and again one would miss its footing and fall into the water. It would then shoot off to the other side like a scalded cat.

Ralph and I and any of the other long-stay campers would wait for our baths until it was dark, then sit in the warm water while Phillipe, Ralph's batman, brought us trays of canapés – smoked mussels on salt biscuits or tuna paste on toast – with a bumper of the 'Strong Dog'. When we could afford it, he would bring an ice-cold bottle of our all-time favourite white wine, the Simonsig Vin Fumé.

They were golden times, those days and nights at the SARA camp at Mlawula.

* * * *

We decided to roast a whole warthog on a spit. Warthog bred quickly at Mlawula and there were tribes of them all over the reserve; they were not only very fecund, they were also very fast, very tough and very cunning. Ralph and Umbuluzi Two had tried before to shoot one of the large tusker males, but they had all proved too elusive. It was a strange thing, when you didn't need to find one, they were everywhere, but if you went out to shoot one, they totally disappeared. Finally, one aging male was shot, dressed and brought up to the camp.

I had watched Christoph bar-b-cueing lambs, but a pensioned-off warthog was another proposition. This one had enormous hind quarters that would take an age to tenderise. We started the fire at six in the morning and put the spit on at eight, to allow twelve hours slow cooking. When we returned at sundown, the aroma of the herbed oils and condiments was mouth-watering, and by eight that night it was ready. I was given a leg to carve, and the meat fell off the bone in neat and fat-free slices. With bar-b-cued onions, roast potatoes and a tomato and apple sauce it was delicious. And there was plenty left over so for the next day's packed lunch we had warthog sandwiches, enlivened by that well known southern African cold meat stand-by, 'Mrs Balls Peach Chutney.'

On that team we had a charming if decidedly insouciant youth called Alvin from Los Angeles. Alvin declared himself entranced by everything in the bush.

"Oh wow, Dave! That's real neat!"

"Hey, Larry! That's really great. No, I mean real great! It's real groovy!"

"Say, Watty, that is so far out, Man!"

But what captivated Alvin the most was a toothsome English beauty named Charlotte who hailed from Britain's stockbroker belt, and she spoke like it too, far back and plummy. She was a fine-jawed, broad-faced debutante with cornflower blue eyes and long blonde hair immaculately drawn off her face and bunched with a clip in a low ponytail at the nape of her neck. She wore light drill shorts and a loose cotton shirt knotted at her bare midriff. A pair of espadrilles on her feet and sunglasses permanently glued to the top of her head completed her meticulously ordered casual ensemble. She was a product of the very best private education Britain had to offer and had then been sent to finishing school in Switzerland. She was now striding out along the primrose path to Cambridge and a degree in modern languages when, all of a sudden, out of life's shrubbery, leapt young Alvin.

Alvin lived with his mother who, following a series of spectacular divorces, had become fabulously wealthy, and as he was the only son he stood to cop the lot. As a result, he had never done a stroke of work in his life. Alvin had never encountered an English rose before and he was at once totally enraptured with Charlie. He carried her backpack for her, fixed her meals for her and, shock-horror, bathed with her late into the night, baboons notwithstanding. When they were excavating on the boards, she with an incurably languid air and he with huge puppy eyes, the conversation was so incongruous it could be quite comical.

"Yes, we always have a family holiday skiing at St Anton every January. Daddy and Mummy have a chalet there, you know."

"Oh wow, Charlie, that's really neat!"

"Then of course we are always in St Tropez every summer. It's super there. Have you been?"

"Hell, no Charlie!" said Alvin, downcast. "We just live in Beverley Hills. Have you heard of Beverley Hills, Charlie?"

The warthog bar-b-cue took place when the two of them were together at Mlawula, so the following day they both enjoyed the delights of the cold meat sandwiches. When they later got back from the excavations, Alvin, revelling in the exotica of Mlawula, turned to Charlie and said:

"Hey, Babe, how d'ya like the lunch? Wasn't that great? I mean, far out! When I next go into our deli, that's what I'm gonna order. I'll say, 'I'll take a double warthog and peach chutney on pumpernickel with a side of fries!' How'd that be, Charlie? You wanna come to my deli?"

Well, she not only went to his deli; unbelievably, they actually got wed. Another marriage made in Mlawula!

※ ※ ※ ※

"So Price Williams, have you found out about these 'Christmas boot flakes' yet?" asked Ralph as we were comparing notes one evening. We had made a good start at Siphiso, but the work was necessarily slow and would get slower as we came to the Late Stone Age strata with, hopefully, all the artefacts to plot and record.

"Not yet," I replied, "but we are slowly getting down to the Stone Age stuff about now. It's getting quite exciting."

"Well, you'd better carry on then!"

So we did.

CHAPTER SEVENTEEN

GETTING TO THE BOTTOM OF IT ALL

If it can be written or thought, it can be filmed.
                                        Stanley Kubrick

Spall, spall and more spall. That was what lay beneath Stratum 2. Spall was like gravel. It represented the tiny spicules of rock that fell from the roof as the cave gently eroded away. It was caused by rainwater migrating through the rock, carrying dissolved salts. The water evaporated as it reached the surface, in this case the ceiling of the cave, and the salts crystallised, forcing the rock to break up and fall like snowflakes onto the floor. There was up to ten centimetres of this abrasive, stony layer, and it played havoc with the enema tubes. There were very few stone artefacts in among the gravel, and they were not in a consistent pattern. From the radiocarbon dates, this spall layer represented almost four thousand years of relative quiescence at Siphiso, a time when the cave was probably used only seldom, if in fact at all.

It was the next stratum, stratum four, which was to reveal an exciting view into the past. As we came to the end of the stony horizons of stratum three, the change in both the colour and the consistency of the deposit was dramatic. Gone was the grey stony gravel, and in its place a series of tiny lenses of light brown soft ash, thick with cultural debris. We were now about six thousand years back in time, just before the period when the first literate civilizations were emerging in Egypt and the Near East.

At Siphiso hunter-gatherer life must have been easy and the shelter was being used often. Among the tools made from chips and chunks of brightly coloured stone were a number of thumb-nail scrapers, so called because of their size and shape. Carefully chipped from colourful raw materials like grey and blue banded agate, or sard, a semi-transparent brown, they looked like little jewels. We found scrapers made from softly-shaded apple-green chrysoprase, orange-highlighted heliotrope, red jasper and vivid orange-yellow carnelian. Today these crystals would be used as gem stones, and maybe the Late Stone Age hunters at Siphiso also had a strong sense of the aesthetic. The crystals came from geodes, solidified glass gas bubbles in the rhyolite, away down the back slopes of the Lubombo. The toolmakers must have ranged far to collect them.

Every scraper we found in stratum four was a high point of the puffing day. Whoops of delight echoed around the cave and work came to a standstill. Everyone wanted to see and touch the new discovery. Once measured, each scraper was cleaned with a small brush for everyone to look at with a hand lens, slowly turning each facet to the sunlight, checking out the working edge to see the wear damage. Each was more beautiful than the last, this one in tasteful heliotrope, that one in shiny carnelian. We all marvelled at the minuteness of the flaking, the incredible skill with which each one had been made. They had been fashioned with an abrupt edge on the one side to clean animal skins for leather production. The hafts in which they had once been held had long since perished, but these artefacts told their own story, of skilled expertise, infinite dexterity and visual satisfaction.

"These scrapers are part of the so-called 'Wilton culture'," Larry explained. "The 'type fossils', the most distinctive tool forms of the age, are found in many parts of southern Africa and date to around six thousand years ago."

We adopted the scraper as one of the SARA emblems. Cindy, one of our volunteers, from Merced in the San Joaquin valley of California, suggested we should start up a newsletter for all volunteers to keep up with the archaeological and social events of SARA. There was an obvious title, *'The Scraper Paper.'* Cindy became its accomplished editor for the life of the expedition. I would send her annual progress reports about new discoveries, as well as who'd married who. Current volunteers would write in with their experiences at Siphiso, or the Long House, or the bush, and past volunteers kept in touch by telling everybody what they had been doing recently. There were lots of requests for Sue's game casserole recipes; 'First catch your impala!'

One curiosity at Mlawula at that time was Brad, from Baltimore, Maryland. He came several years in a row, but most unusually in each case he stayed not for the normal two weeks but for two or three months. He told us he had retired early from the U.S. Navy and was now at a loose end, and that he enjoyed Africa. He had a taciturn air but was good company and did all the jobs he was asked to without complaint. He let it be known that he made something of a living by gambling at weekends. At one of the hotels in the Ezulwini Valley, only a few miles from the Long House, there was a casino and on Saturday evenings, Brad would take himself off there.

Down at Mlawula he would go out with the team every morning and do his stint in the cave, but he frequently complained of back trouble. He thought it must have been all that lying on the boards, he said, so he would ask to be excused for the afternoon. We watched him as he hiked up to the crest of the Lubombo. He would disappear for an hour or two, then reappear with the droppings of various antelope for us to identify just as we were about to return to camp. Scatology was his contribution to understanding the bush, and he talked about it a lot.

It all looked innocent, until one day the game scouts told us they had seen Brad a couple of days during that week near to the Mozambique border fence. At that time the civil war in Mozambique was at its height and the Russians had become deeply embroiled in this post-colonial backwater. Reports from South Africa suggested that a large fleet of Russian warships was at anchor in Maputo harbour. This was right in the middle of the Cold War, so any movement of Russian naval vessels was of considerable interest to the U.S. military. It was then that we put two and two together. Brad was CIA. He must be. He was a retired naval officer, but young enough not to be of retirement age. He came from Baltimore, which was no more than thirty miles from Langley, headquarters of the CIA.

He would wander off and stare at Maputo a mere forty miles away, where the whole Russian fleet, or at least a part of it, was currently holed up. And in any case why did he want to stay with an expedition researching into prehistoric Africa for weeks on end every year when he had no previous connection with the subject. And he said he made his living gambling, except no one had seen him do it. Was he picking up from a dead letter box on the border during the week and contacting a handler in Langley at the weekend?

"Do you think he's a spook?" suggested Colin.

"What are you going to do, Colin?" I replied. "Go up to him and say, 'Hey, you! I think you're a covert agent for the CIA.'"

He did just that. Brad laughed nervously, but then, his laugh was always nervous. As usual, he remained taciturn. We shall never know, but one feature of his activities stands out above all others, his scatology, and it's a fact that the CIA forever deal in crap!

* * * *

"Hi! My name's Nigel. We are thinking of making a documentary for Channel 4 about your work in Swaziland. Actually, it's a series about Oxford really, sort of cutting edge research, big science, that sort of thing."

Making the documentary proved a great diversion, if a bit unsettling. 'Here is someone walking to the site. Here is someone taking his shorts off. Here is someone going to the toilet.' The choreography went on for days at a time, constantly taking and retaking scene after scene. We were fascinated by this exotic team, who hailed from a different world, a world with its own bizarre mannerisms and language.

"Look, David," it was the presenter, Bruce, speaking, "Nigel wants me to do this piece with you to camera, close in. He's already done Andrew. It's 'Why are we here?' 'What are we doing?' 'Why Swaziland?' You know. You're an old pro. Just ham it up for me. And …ACTION!"

"So, David, why are you here? What are you doing? Why Swaziland?"

I made one or two mediocre attempts at an answer.

"Sorry, David, can you give it more glitz, more umph, you know? And smile! SMILE! So, David, Why are you here? What are you doing? Why Swaziland?"

"Well Bruce, only when we have geologists, archaeologists, botanists, geomorphologists and zoologists all working together, only then will we hear the music of science!"

"And …CUT! That was fantastic, David. That's it, you know. 'The music of science,' I love it. Don't you just love that, Nigel? I mean, Christ, you should have your own show. Isn't David just great?"

"But Bruce," I remonstrated, "it doesn't mean anything, 'the music of science'."

"Who gives a flying fuck, David? It sounds fantastic. This is television; it doesn't have to mean anything. OK, is that a wrap for today guys?"

Sometime later we were down at Mlawula where we filmed interesting pieces in the bush, the 'He's an archaeologist, I'm a geographer, why are we working together?' sort of dialogue. The one sequence they really wanted, and which was to be the closing shot of the film, was Andrew and me discussing 'big topics' sitting in the tin baths on top of the cliff as the sun went down. 'Popularising knowledge,' they called it.

On their last night we gave the crew a huge send-off party at the Long House. Toasts were made and boxes of wine were consumed. We all vowed to meet again, for the rushes, for the preview, for the opening night of the series. Bruce was busy thinking up a whole new sequence on African archaeology, or bathing through the ages, or anything. It was a great night. And in the morning they left; we never saw them again. It was as though one night a flock of exotic, colourful birds had flown in through a window of the Long House, twittered and preened for a few moments in the light, and then flown out through another window into the dark. In due course the series was aired, though not to any great acclaim, but we each have our copy of the film, a wonderful reminder of our Hollywood days in Swaziland.

CHAPTER EIGHTEEN

ROYAL GEOGRAPHICAL SOCIETY

There are plenty of problems in the world, and doubtless climate change - or whatever the currently voguish phrase for it all is - certainly is one of them.

P. J. O'Rourke

We were back down in stratum four and the ashes of innumerable hearths. More scrapers were turning up, along with some of the little chips that represented the waste from their manufacture. In a few cases, after a tremendous effort on Larry's part and many man-hours poring over the debris, he was able to fit a few of the flakes back to the cores and the scrapers. This was a very significant discovery, as it was proof that some of the artefacts had actually been manufactured at the site, even though the raw materials had been brought in from some distance away.

So, six thousand years ago, these hunter-gatherers had lived here and made and used their implements here. The scrapers told us that they must have been cleaning and tanning skins which we presumed they used to make other kinds of equipment. From the leather they would have made skin bags to carry their bows, spears, fire-making sticks and digging sticks. They would also have made simple clothing from the treated pelts. Proof of their craftsmanship in leather came with the discovery of a needle with an eye. It had been crafted on a tiny bone from the front leg of a scrub hare. The 'eye' occurs naturally at one end, but the rest of the shaft had been painstakingly filed down, presumably on the rough surface of the nearby rocks, until it was honed to a fine point. They would have drawn thread made from plant fibres through holes punched in the leather to sew the pieces together as garments. There were still fibrous plants growing in the dry streambed, and we were able to show how this had been done.

Another exciting discovery in stratum four was the appearance of jewellery, small disk-like beads made from pieces of broken ostrich eggshell. As a valuable source of food ostrich eggs were very important to the Late Stone Age people. Each female ostrich can lay up to twelve eggs at a time, often in the same nest as another female. It's not uncommon to see twenty or more eggs being incubated by a single male bird. It was easy for these prehistoric people to take a few eggs and still leave lots behind. Incidentally, you can make an omelette for more than twenty people from one egg.

They had other uses too. By carefully extracting the white and the yolk through a small hole in the one end of the egg, the resulting empty shell becomes a serviceable container. They would have used them for carrying and storing water, filling them from pools and springs and carrying them back to their rock shelters and encampments. The Kalahari San went one stage further. They filled the eggs with water in the rainy season and buried them deep in the sand with a straw stuck in the top. During the long, dry winter months, when the ground was parched and hunting involved trekking great distances across arid plains, they were able to search for game over vast ranges because they knew they had a water supply buried at strategic places. They could uncover the top of the straws and sip the water from the eggs without disturbing the whole batch before continuing on their way. When Europeans first watched San hunters disappearing into the waterless desert, they were totally puzzled by their behaviour. Any normal man would have died in those conditions.

We don't know if our Siphiso people cached their water, but they used the shell of the ostrich egg for making beads. They would have found pieces of shell around the nests after the young had hatched. These they broke up into small pieces and drilled a neat hole through the centre of each with a chipped stone borer. Each bead would have been shaped into a rough circle and threaded onto a sinew lace. When they were all assembled, the whole line would have been neatened by rubbing the string of beads in a stone groove until they were all the same size. They would then have been rethreaded into intricate and ornamental patterns – headbands, necklaces, bracelets and pendants. Such decorations are portrayed on some of the human figures in the rock paintings in southern Africa. The beads and pendants probably carried a symbolic meaning, such as announcing which family they belonged to and whether they were married or not, a sort of personal identity. If so, then our beads at Siphiso were the marker of a complex society back in the Stone Age.

* * * *

As the digging season advanced and the rainless winter wore on the bush was becoming drier and drier. Slowly, the pools in the stream bed below the cave were shrinking, and the frogs that had made their home there became more vulnerable. By August the marabou storks had arrived for the opening of the fishing season. They gathered in flocks along the river banks, huge, ugly, bare-necked and bare-headed birds that stood like undertakers in black frock coats. With them came the even larger but less cadaverous saddle-billed storks, with long, striding, stately walk and brightly-coloured red and yellow bills ringed with a central black band. They were the priests officiating at the forthcoming funeral. When the water had fallen to a critical level, the service began. With keen eyes and rapier bills the saddle-billed storks speared their prey and with a jerking adjustment of their beak, swallowed the frogs whole. The marabou did the same, stabbing from side to side, engorging the unsightly skin pouch that hung down in front.

Occasionally, fish eagles would fly in from the Mbuluzi gorge and sit high in the knobthorn branches waiting their turn. They were beautiful birds, tawny breasted with dark brown wing feathers, a striking white chest and head and a bright yellow bill. Their sharp, ringing cry carried far over the bush; it is one of the great sounds of Africa. They came into Mlawula to feed on the barbel that had become marooned in the upper streams as the water dwindled.

The rainless weeks continued. Even the dry grass stalks were gone now, chewed down to the very roots, and most trees had lost their leaves. They stood, sterile and bare-branched against the pale sky. The impala and wildebeest assumed a baleful and lean look. Those that had not built up enough reserves of fat in the green summer began to weaken from lack of food. As the feeble ones stumbled, the vultures gathered to await the end. Day after day we watched the slow inevitable decline; the closing drama would not be long in coming. Each morning brought with it a whiff of carrion on the chill dawn air. Another animal had succumbed in the night, starved of grass and nipped by cold. By the time we found it, the vultures were busily devouring the last of the flesh, a seething mass of fighting, squealing birds hacking and pulling at the innards with their heavy, hooked bills. Despite the tragedy it represented, it had its quirky side too. These white-backed vultures often spent more time running at each other with their clumsy, hopping jumps, trying to keep others from the carcase, than they did actually feeding. If a pair of black-backed jackals joined in you could begin to offer odds as to who would get any meat at all.

The bush was white with dryness. Only the occasional green-leaved fig offered a distant remembrance of rain. The wild gardenias had long since flowered and only the hard egg-shaped fruits adorned their once sweet-scented branches. For the baboons the plump fruits of summer had gone long ago. They sat disconsolately on their haunches among shafts of slanting sunlight in the dry water courses, turning over the stones to look for beetles and spiders. They would even eat scorpions if they could find them, gingerly grabbing the back of the tail, snapping off the sting, and, as if eating oysters, tilting their head back and dropping each one into their open throat. Even the rhinos meandered about listlessly. With no mud to wallow in and little food to eat they too were waiting for the rain, and the heat. Every year there was a sense that this time the drought was going to be permanent, that somehow the rains would fail for good, and that verdure would never return to the bushveld. The leafless branches of the marula trees stuck out like accusing fingers pointing at the cloudless heavens and the last desiccated foliage of the raisin bushes fell silently to the floor, exposing their nakedness. It was a time of unending deprivation.

But then, just as the bush was giving up its essence to infertility, a small miracle occurred at Mlawula. The knobthorns bloomed. At first only tiny red buds appeared among the dehydrated twigs, then one by one they opened into a showy display of long, delicate, creamy catkins, so that the air became filled with a most pungent mimosa fragrance. One by one each tree gradually came into flower, a mass of soft, off-white blossom that transformed the veld. It was not the rain. It was not even the harbinger of rain. But it seemed to lift the bush in a spirit of expectation. Rain could not be long in coming. I wondered how the Stone Age people must have felt as they looked up and saw the same sweet-smelling snowy flowers. The fragrance of the knobthorns must have told them that the seasons were turning, that soon the year would come full circle, that presently the storms would gather and fatness would return to the land. Like the trees and the animals, the birds and the insects, the Stone Age people were used to the drought. They had been brought up with it. They had evolved with it. They had learnt the secret of dearth and plenty and had survived millennia after millennia. They had become infinitely skilled in the art of staying alive, in harsh times and in times of abundance. Only intruders from the world outside would ever destroy that achievement, and they were not due for thousands of years.

* * * *

Stratum four petered out, exposing stratum five, slightly coarser but otherwise containing much the same material, only in greater quantity. We were back almost seven and a half thousand years ago, according to the radiocarbon dates. There were more scrapers, more flakes, and a rich collection of bone pieces to deal with. Enter our paleo-zoologist, Dr Jones, not unnaturally known as 'Bones Jones' whose job was to identify the bone fragments that represented the remains of meals our scraper people had enjoyed, to try to work out what their hunting strategy was, and to see if there was any change in the environment this long time ago. We were uncovering all manner of bones, lots of toe bones, lots of ankle and leg bones, and nearly all from small antelope. There were hardly any ribs or vertebrae or skull bones.

It became clear that animals were being hunted some distance away, either by bow and arrow, or by snaring, and that after they had been dismembered only the meaty limbs were carried back to the shelter. This would account for the high incidence of toe and ankle bones, and the lack of back bones or crania. There were also lots of scrub hares and mongoose bones. These would have been caught by trapping or snaring; the fact they were small and predominant among the food remains suggested that the people living at Siphiso at that time had been a small family group. Other bones that turned up belonged to klipspringer, small antelope that still hop about on the rocks above the cave today, and reedbuck, which have been seen on top of the Lubombo. Apart from a couple of zebra hooves, which had probably been picked up from a kill elsewhere to use as anvils in bead-making, large mammals were missing from the record, again suggesting a small family unit.

We found the bones of tortoise, easily caught and cooked; the bones were often charred. Also there were pieces of broken and charred snail shell. These were not small European-sized escargot but enormous African land snails that can grow up to twelve centimetres or more; they are still to be found in the bush. The families at Siphiso were heavily into snails, but where, you wonder, did they get their garlic butter?

It was very satisfying to be able to reconstruct a day in the life of these people almost eight thousand years ago. They enjoyed a varied diet of hare and tortoise, snails, leg of venison, ostrich-egg omelette and a spot of pork from warthog. From the mass of marula pips we found they must have gorged themselves on these sweet apricot-like fruits every autumn. There would also have been figs and raisins, melons and squashes, catfish and land crabs - a veritable prehistoric cornucopia. The families that lived here, probably for a few months each winter, made clothes here and decorated their bodies with beads. From the slivers of red iron oxide we found they made paint too, to paint themselves? Or paint their cave? We shall never know. But one thing was clear. The iron pigment came from the high mountain ranges to the west. Did they travel all that way, one hundred kilometres towards the land of the setting sun, or did they barter for it from other groups living in the higher lands beyond the bush?

Our digging season was coming to an end. The next layers would have to wait until the following year.

* * * *

"I think," said Andrew, "that it would be a really good idea to put together a group of members from the Royal Geographical Society and bring them out to Swaziland to show them 'geography in action.'"

Once the volunteer diggers had left, a group of these august people duly arrived at Matsapha Airport and we drove them up to the Long House for the first few days of their stay. They were an impressive company - a poet, a geologist, a criminal lawyer, a high court judge, a TV producer, an astrologer, a captain of industry, a couple of school teachers and a few retired civil servants.

Having checked in at the Long House, we went out at once in the Land Rovers for a couple of hours' orientation up the Mdzimba mountains where we could look out across the middleveld. At the highest point of the Mdzimba, the view is spectacular. Looking to the one side the mountains of the highveld fill the western horizon. In front the land falls away into that great scuttle of the Ezuluwini valley, and then rises on the other side to the crags of Nyonyane, above the Long House, and then up higher to the unmistakable twin profile of Sheba's Breasts. I explained that they had all been formed at different ages and from different chemistries of granite, all exceedingly old, older than any rocks they may have seen before. I took great care to get my facts correct knowing that in the audience we had a geologist, Bill, who was the curator of a geological museum.

We got back to the Long House to discover that in our zeal to get people into a geographical mode we had failed to notice we had left two ladies behind. I was confronted by one of the pair as we walked in through the door.

"Well, really, I think it's a disgrace! I don't mind for myself, you must understand. But you left my friend Bunty towering with rage in the shower!"

I was deeply embarrassed, and with that, Bunty appeared in person, the epitome of mild manners, who apologised for being delayed. I took them on a private trip back up the Mdzimba. Honour was satisfied.

The next day we visited the highveld and Ngwenya, the iron ore mine and Lion Cavern. As we walked across the summit of Ngwenya I described the geology again, the oldest sedimentary rocks in the world, and the blue-green algae, the Earth's oldest living life forms. At each stop, I selected hand specimens and passed them to Bill, who appeared to put each one in a bag. I learnt later that what he actually did, when my back was turned, was to toss them over his shoulder.

"I didn't come here to look at bloody rocks!" he confided to a fellow traveller. "I get too many of those back home."

We continued with our walk.

"Oh dear," said Bunty, "I've dropped the little screw out of my glasses!"

The ground was littered a foot deep in Middle Stone Age debris. Everyone tried to help, creeping around lifting stones. The astrologer came to their rescue.

"Do allow me to assist you. I have this astrological talisman that picks up the cosmic vibrations from lost objects. I use it a lot at home in Stoke Poges."

She pulled out a carved amber pendant in a heavy silver setting suspended by a fine chain and began twirling it across the ground.

"It's this way!" she crowed. "Over here!"

Half a dozen men moved on their hands and knees in the direction indicated.

"More to the right!"

The pendant whirled right, then left, now faster, now slower. The six men crawled first right then left until eventually group embarrassment set in and they were encouraged to give up. I assured them that we would be able to find another screw anyway at the optician's in Mbabane.

"Well," said the astrologer, "it's never failed before. Maybe it's because we are in the southern hemisphere and the lines of cosmic force rotate in the reverse direction. Yes, that will be it. Africa must be a bit different to Buckinghamshire!"

The next day was a long drive through the middleveld, across to the Komati, then up to see some of the rock paintings. The poet recited an ode to Insangweni.

"The trance takes hold," he pronounced, "but the imagined wings do not bear us up."

Sue and I had moved out of our bedroom in the Long House for the duration of the tour and were staying with Christoph and the family. This allowed the TV producer, Storm, and her husband Robert, to have a double bed to themselves, which they had specifically asked for. We'd had the double bed specially made for us from local pine and the mattress was comfortably supported on lathes. The next morning when we returned to the Long House before breakfast, to our astonishment we found that half the soft coverings from the arm chairs had disappeared until Storm came into the common room clutching the missing cushions.

"It's the slats!" she said by way of explanation. "They're getting to Robert!"

There was no answer to that, except that when Toko asked why Robert was wandering around groaning and clutching his back, I said,

"Don't worry, Toko. It's just a slight case of the slats!"

Storm and Robert wore matching moss green Rohan travel wear, the ones that have the 'On Route' jackets and 'Globetrotter' trousers covered in zips, which were all the rage at the time. Storm cornered me one day after we had given a slide show about the dongas.

"Oh such a pity you don't use helicopters! They are so convenient for that sort of thing. We use them all the time in the television industry. You should really think of getting a couple, you know; they'd be so handy in your line of work!"

The visit ended without further incident.

* * * *

Chocolate, that's what the volunteers talked about at coffee breaks under the marula tree on the rock at Siphiso - chocolate and ice cream. They drooled about this store in St Louis that sold the best caramel and chocolate ice cream with marshmallow whirls, or that store in San Francisco that made cherry and walnut ice cream with chocolatey chunks. It probably helped to alleviate the slow hours spent stretched out on the boards in the cave, head down in the dust of the ever-deepening excavation.

We were beginning our last season at Mlawula, when the final secrets of Siphiso would be revealed. In stratum six and seven, the deposit became much coarser and the tools, instead of being made with the fine glinting chalcedony, were being made of the dull rhyolite of the Lubombo crags. The flakes were big, almost awkward by comparison to the refined scrapers and flakes of the later periods. But there, in one square after another, first one, then the next, 'Christmas boot' flakes began to appear. We had finally tracked them down, the 'Christmas boot flake people.' When we got the dating results back from the laboratory, we established that they had lived in the Nkumbane Valley more than nine thousand years ago. They predated the scraper people. Why had they made these coarse tools? Why hadn't they found the agate and carnelian and used that instead? One of the volunteers made the obvious inference that they must have been a lot less sophisticated.

At first sight that is what it seemed like. We had become so used to the beautifully made gem-like tools in the upper levels that these older flakes did appear crude and clumsy. But then again, this rougher type of technique littered the bushveld. It was everywhere, so these people too must have lived successfully at Mlawula. Could we really use the quality of rock they chose, or the fineness by which they made their tools, to define how advanced or backward each group was? The answer had to be no. They made the tools they needed to survive; they hadn't made them for the delight of archaeologists. So if that was not the reason for the big flakes, what was?

The more we examined the 'Christmas boot' flakes the cleverer they were. They were not just slabs of rock whacked thoughtlessly from the rhyolite outcrops; they had been shaped, chip by chip, on a complicated core and then the pre-shaped flake carefully removed. We found the cores in the cave, and we also found some of their quarries outside down the valley. One of their remarkable characteristics was that they had been made with one final oblique blow across the working surface that had created a sharp, diagonally flanged edge like a chisel, the toe of the boot as it were. They looked for all the world like wedges. Was that what they were? Wedges? Wedges for working in wood? For removing bark from trees, or splitting timber? Was the climate they lived in different from the scraper people so that they were using bark instead of leather? Were we seeing the difference between a woodland culture as opposed to the later more open savannah people with the scrapers?

We can never be precisely sure, but other evidence was beginning to point that way. In the bone remains of the scraper people we had found evidence for Cape fox. This lovely animal doesn't live in this part of the world now. It is found only in the more arid edges of the Kalahari in Botswana and the perimeter of the Namib desert. Supposing the climate had been drier five or six thousand years ago, after what had been a wet phase earlier than that. Could that have accounted for a change from working in wood to working in leather, and so the change in the tools people were making? If the tools were telling us that, then the sediments in the cave seemed to be suggesting that too. We had found lots of spall with the 'Christmas boot' flakes yet hardly any with the chalcedony scrapers. The more rain, the more seepage; the more salts; the more weathering; the more spall. The charcoal evidence was to point the same way too. And what about the huge gap between the scraper people and the pottery users, when Siphiso lay unused from six thousand to only two thousand years ago? Was that four thousand year break in occupation the result of a time when it became drier still, when not even the accomplished scraper people wanted to live here in the bushveld? It certainly looked that way. There had been a long and progressively dry spell in the second half of the last ten thousand years. That is what had happened. And the Stone Age people had responded accordingly. Whenever the climate became impossible, they moved away.

Underneath the 'Christmas boot' levels we came, low in the deposit, to another surprising discovery - the tiny microblades we had seen in the open air sites like the rhino pan. Their use will always puzzle us. They have been found in the Cape of Good Hope at around the same time. In our strata they turned up in eight and nine, dated between eleven and twelve thousand years ago. What were they for? No one seems to know – arrow head edges, maybe - but they are very different to tools in the later levels, and they date to a time when the world's climates were changing dramatically.

More importantly, beneath the microblades there was nothing. The deposit was still there, but this lowest level of gravel was totally sterile. There was no Middle Stone Age at Siphiso. In fact it was not common anywhere in the bushveld. Was this because the climate made it untenable too? What were we to make of these waves of occupation then oceans of silence? Our assumption is that we were seeing human responses to great oscillations of climate.

※ ※ ※ ※

It was then that the jigsaw fitted together. The evidence from the dongas of the middleveld was finally being understood; the organ-pipe sediments from the Mkhondvo Valley and from sites as far away as the Zambezi in the north, and south into Zululand, were giving up their secrets. We had analysed these in every conceivable way. We had looked at the surface of individual grains with a scanning electron microscope to find out if they had been rolled in water, or been battered by the wind, or if they had come from an earlier river deposit. The micrographic pictures of these grains were amazing. The surface of each tiny grain showed up like a cliff face where every record of its history was clearly written.

We had looked at the sediment chemistry, at all its physical properties, at its distribution, elevation, latitude, longitude. We had now even obtained radiocarbon dates for it. It had been laid down between thirty and twelve thousand years ago. And what did it mean, the sediment we called colluvium? It meant that between those dates much of southeast Africa was arid. It had been a dry land, perhaps even a semi-desert. We found supporting evidence all over southern Africa – old sand dunes, ancient dry lakes, and this organ-pipe sediment, deposited by periodic flash floods around twenty thousand years ago in what for humans was a less than hospitable world.

I had been brought up on top of a glacial moraine in the Marcher country of Britain at latitude fifty three degrees north. The house stood on an enormous heap of ancient glacial outwash from the heart of the great ice sheet that had covered northern Britain and much of Europe twenty thousand years ago. This was the height of the last Ice Age, when our world was going through the coldest time in recent Earth history, a time of great privation, and even greater uncertainty.

But what had happened in the sub-tropics during that globally cold phase, here in Africa at twenty six degrees south? For decades scientists had suggested that the climates of these equatorial and mid-latitudes hadn't changed at all, that they were unaffected. But now we knew that they had changed; they had been as strongly if differently affected, meaning the climatic cycle was worldwide.

The answer we had obtained from Swaziland to the question of what happens during an Ice Age was, when the world gets cold, it gets dry. More than ten years it had taken us to discover that result; ten years to define that simple sentence. Our colleagues in Australia and South America were finding the same thing. For whatever reason, when the Earth gets cold and ice creeps out from the poles, the tropics and sub-tropics, although unaffected by ice, become dry, in some cases hyper-arid. Wind-blown dunes had been found under the jungles of the Amazon and desert sands under the rain forests of the Congo basin. One of our own team in southern Africa had defined the Kalahari a mega-desert during this time. Today it's only a fossil desert; indeed some parts of the Kalahari today get more rain than London. But during the ice ages, the Kalahari was one vast, totally arid, central-continental desolation. So, Africa did react to global cooling.

Most importantly, it was in Africa that mankind developed, against this ever-changing backdrop. There have been at least twenty of these cooling events in the last couple of million years. What a bewildering array of climates our ancestors had to contend with. What staggering changes in environment they had to overcome, from blood-curdling cold to warm easy days; and from searing, barren desert to tropical plenty. It is amazing they had survived at all. Yet they did, by evolving, culturally and physically, and by inventing ever more successful technologies and ever more specialised life ways. They survived first by understanding and then slowly by dominating their environment to become 'us', the ultimate in environmental specialisation and dominance. We are the end product of this evolutionary process. But without our prehistoric forebears' growing comprehension of the world around them in the Stone Age past, with all its changes and challenges, we wouldn't be here at all. From the long-gone glaciers of my birthplace to the ancient deserts of Africa, my experiences had come full circle, encompassing the globe.

And what about the Stone Age people of Swaziland? What had happened to them? In the later stages of the Stone Age all was becoming distinct. At the height of the last Desert Age - and that is what we would have called the ice ages if we had started our research from Africa, instead of from of East Anglia or Wisconsin - some twenty thousand years ago the bushveld was empty of people. It was too dry. The cold Indian ocean and the rain shadow of the Lubombo conspired to render it uninhabitable. Small groups of Middle Stone Age people lived in the middleveld dry valleys, where the colluvium was forming, but that was it. If the evidence from Sibebe was anything to go by, they had even abandoned the highveld as well; too cold perhaps. The Middle Stone Age tools up there were earlier, followed by a long sterile time.

Then, about fourteen thousand years ago, world climates started to warm up. The ice sheets at the poles melted. Sea levels rose by up to one hundred and fifty metres. Ocean surface temperatures climbed. Rainfall increased. In southern Africa the deserts retreated and savannah returned to the bushveld. People started to live at Siphiso and the Nkumbane valley, probably after an absence of thousands of years. The microblade makers, then the Christmas boot flake makers, then the scraper makers.

The climate around six thousand years ago seems to have dried out again, not to the same extent as the earlier time of extreme drought, but enough to slow and then halt occupation at Siphiso, caused by a small climatic fluctuation in the Holocene, the most recent geological period. By one thousand five hundred years ago, the rich savannah returned yet again, but it was too late. The agricultural invasion drove the Stone Age people to extinction in Swaziland, and now the only relatives they had left were dwindling away in the Kalahari, the last of the Bushmen of southern Africa. In their place the cultures of the African Iron Age had developed, culminating in the emergence of Swazi society, with its complex culture and customs. That was the full story.

* * * *

"So, I understand you're an archaeologist. How absolutely fascinating! What an interesting career you must've had. My goodness! All those mummies in Egypt! All that gold from Troy! Stonehenge. Ephesus. Fantastic! And what's your most interesting discovery? You know, the find which gave you the biggest thrill?"

"When it gets cold, it gets dry."

"Sorry, I didn't catch that! What was that?"

"That's it; the most interesting discovery of my life. That's what we found. When the Earth gets cold, the sub-tropics get dry. It's as simple as that."

Time and again at parties the same thing happens. I can see the wheels going round, the eyes glazing over. What good is that? But the answer couldn't be clearer. Human beings are always at the mercy of Earth's ever-changing climates. They have adapted and improved their strategies for life around fluctuations of temperature and rainfall. Even our own ultra-modern urban civilization is rooted in the soil of climatic change. And we know that the Earth's climates never stay stable; something will happen again and probably soon, whether we contribute to that change or not. It always does, and when it does, we shall not escape its effects; we may even be wiped out by it. Global cooling, global warming, climatic catastrophe, we have lived through them all; in fact, they have so far been the engines of our physical and cultural evolution. But will we get through the next one?

Mankind began in Africa. Our forebears took their first steps on this continent five million years ago, probably because the climate became cooler and drier. Grassland appeared at the expense of forest and some of our once tree-dwelling ape cousins started walking on two legs for the first time in this new open landscape. There was plenty of fossil evidence for that change from sites in nearby South Africa.

Two million or more years ago it was in Africa too that our ancestors, using their newly liberated fore-limbs and hands, developed the first technology, the first stone tools, like the ones in the Mzimnene gravels of southern Swaziland. Chipping away at the river cobbles to form choppers they were able to compete in the ever-growing open savannah. Among the grassland animals and their predators that was our ancestors' answer to their own lack of teeth and claws; make them! And with them, and with the later more advanced hand axes, they worked out new strategies for survival. They kept ahead of their competitors by using their brains and by conceiving and making the necessary equipment. And with larger brains and the relevant implements our forefathers expanded, eventually migrating from Africa into Europe and into Asia, stepping outside their environmental niche for the first time ever. No mammal had ever done that before, not so absolutely.

By the Middle Stone Age in Africa they had developed complex tools, like the ones at Ngwenya and Sibebe. In this respect, from the new dates we have, Africa was still way ahead of Europe. Africa remained the continent of innovation and invention long before these same improvements came to Europe. Crucially, Africa is not, as had been thought, peripheral to our development; it is in fact paramount. Many would say that even contemporary humans, truly the 'us people,' represent yet another cultural and physical enhancement which began in Africa. That's what the DNA evidence is now suggesting, that we ourselves all have recent African origins, yet another, if not the ultimate, advance from the great continent less than sixty thousand years ago.

Then, as little as ten thousand years ago, this time in the Nile valley of northern Africa, people initiated an economic paradigm-shift which was to have stratospherically far-reaching consequences. They developed agriculture, the domestication of plants and animals, itself a response to changes in the Mediterranean climate and an increase in rainfall at the close of the last ice age, as we had also discovered in Gaza. They moved from hunter-gathering to settled village life, thence from village to city and from the Stone Age to the ages of metal. During the last quarter of one per cent of global human time they developed the first literate civilizations on Earth. Through total manipulation of their environment, contemporary humans encompassed monumental architecture, metalworking, pottery, writing, and the whole societal panoply of the modern world. It was from here, and because of what they had learnt in the Nile valley, that the black races migrated south across the equator into southern Africa with the whole cultural and economic package of the Iron Age - cereal crops, cattle keeping, settlement, metal working and the complex rituals of sedentary living. It was that plant-growing, cattle-keeping imperative that impelled them to replace the age-old world of the Stone Age San.

Africa has it all, the whole history of mankind. *'Ex Africa semper aliquid novi,'* as the Roman author Pliny wrote, 'There is always something new coming out of Africa'. He could not possibly have known just how phenomenally right his statement was!

And in Swaziland we had that whole story, right up to the last of the great traditional African monarchies. What a story! What a span! The people of Swaziland ought to know what an illustrious and extensive heritage they have.

And that's how I got involved with building the Swaziland National Museum.

CHAPTER NINETEEN

EVOLUTION AND THE METHODISTS

Did you ever notice how people who believe in Creationism look really unevolved? "I believe God created me in one day!" True; in your case it looks like he rushed it.

Bill Hicks

The 'museum' at Lobamba resembled nothing so much as a forlorn, vacant locomotive shed in a muddy field. Johnny asked me if there was anything I could do to help, to try to organise some kind of exhibit, anything to fulfil the National Trust Commission's remit of preparing a museum to display the nation's heritage.

The person nominally in charge of the empty shed was Shadrach, who, without any support from government, had attained a state of deep-frozen paranoia. I would move two steps forward, and Shadrach would retreat three steps. We needed re-enforcements, and they came in the person of Sandy, the wife of the local manager of South African Breweries. Sandy was interested in interior design and was attracted to the grass-work made by the Swazi women - the mats, the beer strainers, the grass bowls and jars. Together we developed a plan to put some simple exhibits into the engine shed, to get the museum working. We divided the building down the middle, half for archaeology and half for Swazi ethnography. We had hardly any budget, but we had great enthusiasm. There were four bays in each half of the building, which naturally fell for us, the archaeological section, into four exhibits – Early Man in Africa, The Stone Age, The Rock Paintings, and the Coming of Farming, which signalled the arrival of the Bantu-speaking peoples of southern Africa. This led on logically to the history of the AmaSwazi and all the ethnographic material that Sandy had collected.

"We need those boxes painted blue, Shadrach," ordered Sandy.

"It is not possible."

"Why on earth not?"

"We do not have the paint brushes."

"Well, go and get some from Mbabane."

"I am sorry Sandy. We cannot go there."

"Why? The Museum truck is outside."

"We do not have any petrol."

"Oh, for God's sake, here's the money for petrol."

"But our driver is off sick today."

"Well, bring some brushes yourself from Mbabane tomorrow."

"But tomorrow I have to go to a funeral. It is my grandfather. He has just passed away. It was a few minutes ago, actually."

We used to compare notes on Shadrach's inactivity, wondering what fanciful excuses he would come up with next. In the following few months, Shadrach's entire family was systematically exterminated, some of them several times, to support a spectacular spate of Shadrach's absenteeism. They apparently invariably chose to pass through the curtain on Fridays or Mondays, allowing Shadrach to extend his weekends. His mother breathed her last at least three times over this period, each death more gruesome that the last. It seemed that head-on collisions were her favourite way of departing this life.

Without Shadrach, we decided we could cope on our own, and that we should produce four dioramas, four snapshots of the past. The first, Early Man in Africa, consisted of a map of Africa showing where Olduvai Gorge was, and the sites in South Africa where fossils of our early ancestors had been found. We obtained casts of some of the skulls, and although we didn't have this kind of evidence in Swaziland, only three hundred miles away near Johannesburg was the richest fossil site in the world where the earliest human ancestors had been discovered dating to more than 2.5 million years ago. By implication, these proto-human life forms would have been in Swaziland as well.

It was then that Watty exhibited his remarkable talent for painting. He copied a series of pictures from a text book, showing a row of evolving humans, from an ape to the first stooping early human, then a more erect individual until at the end was a striding figure representing modern man. To get himself into the artistic mood, Watty would creep around the Long House in a crouching walk, pursing his lips and uttering ape-like cries of 'Oo! Oo! Oo!' The kids picked that up and soon the whole house was full of chimpanzees.

The second bay was devoted to the Stone Age. This was a difficult concept to portray to an audience that has no background in prehistory. For Swaziland's earliest inhabitants we displayed a few choppers from the gravels of the Umzimnene valley, and some hand-axes from the Ngwempisi. Then somehow we had to show the Middle and Late Stone Age. We decided to recreate Sibebe Shelter, with photos of the site, a reconstruction of the various levels of the excavation, and then exhibiting the implements being used. We fastened the projectile points onto wooden shafts, and the scrapers we put with leather skins. We built special boxes that would light up the translucent stone implements like jewels. We chose the most colourful spear points and exhibited them in highlight and shade, 'Jack's Point' being the centrepiece. We passed light through Late Stone Age quartz scrapers on an otherwise black background so they shone like diamonds.

In the third bay Watty recreated the main scenes from Nsangweni Rock Shelter, including the four dancing men mincing across the picture holding bundles of twigs. Below these were the floating shamans in trance. More cries of 'Oo! Oo! Oo!' were heard as Watty got into dance and trance mode. His copy was very like the original, and beneath the backdrop we added the various colours such as the iron ore from Ngwenya to show how the paintings had been made. Our kids were fascinated by Watty's skills, and his ape-like antics. Ever afterwards, whenever our children saw a new rock painting with human figures, they would shout out, "Look Dad . . . Oo-Oo men!"

The last bay showed a map of the coming of the Iron Age, how the black races of Africa had migrated with their newly-domesticated cereal crops like sorghum and millet from north of the Equator into southern Africa. We cut out huge curved wooden arrows to show the direction of travel, around the rain forest into Angola and Namibia, and down through the highlands of Tanzania and Malawi, across the Limpopo into southern Africa.

We added some clay pottery we had excavated at Siphiso and a small iron-smelting furnace we had found in the Mkhondvo Valley, complete with slag and tuyères, the ceramic tubes through which the air was blown with bellows and we even got Johnny to part with a pair of traditional bellows to show how it was done. On the wall next to the furnace we exhibited a selection of assegai points. Interestingly enough, the great Zulu warlord Shaka had sent his indunas to the AmaSwazi ironsmiths who then made for him his unique broad-bladed assegais with which he carved out his empire in the early years of the 19th century. Maybe this furnace from the Mkhondvo was one of the ones they used.

We had some difficulty titling the exhibits. The English was easy but the siSwati proved elusive. We had a young lad named Boy Mhlanga who helped out with odd jobs and I asked him how to translate the various labels. There was a slight misunderstanding one morning when I called over to him.

"Boy, what happened to that tea you were making?" I shouted.

One of our volunteers from the USA took exception to this and accused me of being patronising with Mhlanga until I pointed out that 'Boy' was his Christian name. Anyway, Boy did well on most of the easy titles like 'stone axe' and 'iron spear.' On the more complicated ones we got stuck.

"Stone Age artists, Boy. How do we do that?"

"*Hawu*! Dokotela. Stone Age artists! That is very, very difficult for me. They are three different things, you know, 'stone,' 'age' and 'artist.'"

"Well, Boy, how about Bushmen painters?"

"*Hawu*! Dokotela, that is very, very difficult for me."

In the end we got it. '*Tichwe leti Badvwebi*.' 'Bushmen that paint,' that was it. I ordered the letters for the titling but before they arrived I showed the translation to our Swazi cleaner.

He looked at the paper for a long time, then screwing up his face he shouted, "Twaaarf! Twaarf! Twaarf!"

"What's he saying Boy?"

"*Hawu*! Dokotela. He is saying 'Twaaarf! Twaarf! Twaarf!'"

"Boy, I heard him, but what does it mean?"

The two men talked together and there was a lot of arm waving and long stares into the distance. Finally, they both agreed. '*Yebo, kunjalo*!' 'Yes, that's it!'

"He means 'dwarf,' Dokotela. The words on the paper, he says they mean 'dwarf fishermen.'"

"But you said they meant 'Bushmen painters.' Whatever happened to that?"

"*Hawu!* Dokotela. I am believing that they can mean both. Bushmen painters and dwarf fishermen, it's the same I think."

When the letters arrived I stuck them on the title board with super glue. Someone else came into the hall and looking at the 'Bushmen painters' titling, put his head on one side and frowning asked, "Dokotela, what is this thing here, 'painted badgers'? Do we have those in Swaziland?"

* * * *

Sandy completed her four bays on time. There was one on traditional dress, one on grass weaving, one for wooden objects like milking stools, and one with leather shields showing the different age regiments of the king's warriors. We opened the museum to the public with little ceremony. I had finished my job of displaying the rich heritage of Swaziland.

However, events took another turn. The Trust Commission was so impressed with the museum that they quickly abandoned their office in Mbabane in favour of new ones built next to the museum at Lobamba. Sometime later I was called into a meeting of the full commission. The commissioners eyed me smilingly. They came from all over the kingdom, elders from some of the great families of the nation. But something was about to be announced and Matsebula launched into a speech.

"Dokotela, you are '*Lumfundzisi lamatje,*' the man who can teach us about stones. But you are also '*Lumfundzisi lamsamo,*' the man that can teach us about museums. '*Umsamo,*' you know, is that place at the back of the *indlunkhulu*, the great house, where the Swazi families store their relics, the sacred belongings of their ancestors. We think that would be the right name for our collections, 'Umsamo Wesive,' the Swaziland National Museum."

Johnny, who was also there, continued, "Everyone is so impressed by the small beginnings we think it is now time for a full-scale museum. We have got the finance from government and we would like to start right away. The commission would like you to manage the whole programme, the design, planning, estimates, building and finishing of a national museum that everyone can be proud of, a place where the treasures of the nation can be stored. The king's diamond jubilee is coming up in a few months, and we would like it ready by then. It will be the Trust Commission's contribution to the celebrations. You will get the details later. Off you go, now."

I met Ralph outside and told him anxiously what I had just been asked to do.

"Well, Price Williams, if you will get yourself into these things, you'd better get on with it! No good just talking about it, you know."

So I did. I approached an architect I knew in Mbabane, Richard Stone, who was known endearingly by the Afrikaans, Dick Klip. He used a whole new jargon of terms I knew very little about. I gave him an idea of the budget, and he worked from there. But I told him we needed it ready in four months' time. He didn't blink. If we had the money, we could begin.

In a week he had drawn up the plans, elevations and specifications. A local building firm who did odd jobs at Mlawula gave an estimate and we began. They moved onto the site immediately with their equipment, including their all-important '*igandaganda*,' the onomatopoeic Fanakalo word for a cement mixer. I acted as glorified clerk of works, overseeing the building as it went up, block by block. There were innumerable small problems, especially with supplies, but the new museum began to emerge from the rubble. One problem I had was screwing the money out of the Trust Commission's accountant, Shongwe, who seemed reluctant to part with the cash, as though he was being forced to pay from his own pocket. But some months before, Shongwe had sidled up to me conspiratorially with a most peculiar request.

"Dokotela, when you are next in Britain, can you please get some video tapes for me? I need them very, very badly."

What was coming next – football matches? Filthy pictures?

"Dokotela, I need these videos. I have written the names down."

I read, 'BBC Ballroom Dancing Championships 1978; BBC Ballroom Dancing Championships 1979."

"What on earth do you want these for, Shongwe?"

"Dokotela, I am the leader of the Manzini Rumba Kings and Queens Formation Dancing Team. We need the tapes to practice."

I bought them but he declined to pay for them, and I forgot to ask again. However, now, when I desperately needed cheques signed to continue the building, I placed the bills on his desk and said, "Come along, Shongwe, Cha Cha Cha! old chap, Cha Cha Cha!"

The cheques were duly signed.

* * * *

"See those two guys building that wall?" the builder asked. "They are both murderers. Did you know that?"

One was wielding a large sledge hammer to break up some foundations.

"Yeah! That guy on the left has killed two people. He was a bouncer at a local casino and took his job a bit too seriously. Anyway he's out on parole at the moment, but what a fantastic worker eh?"

I gave him a wide berth. Another two bricklayers were building a parapet in front of the old 'engine shed', to disguise its profile. They had put up scaffolding and were cementing in rows of concrete blocks across the front. It was going higher and higher. The builder was away the next day, but no one told his labourers to stop so they raised the scaffolding more and more and the wall continued up and up. By the time the builder came back, they were way up in the air. They spent an unhappy two days pulling it all down again.

It was time for the roof. A specialist engineer was called in to make all the steel trusses. The girders looked as if they were rejects from the Forth Bridge; they were gigantic. I asked why we needed such huge metalwork. Dick Klip said that the roof would be able to hold a ten-ton truck driving over it; we never knew when we might want to do that, he said. He also mentioned that at some point during the next few days they would want to throw the concrete floor. He explained that because of a membrane beneath the floor, it would all have to be thrown at the same time, a mammoth job that would take fourteen hours. Clearly the *igandaganda* would be working overtime. Once completed nothing must ever be allowed to disturb the slab.

I remembered that I had asked for electric sockets to be placed at points in the floor in the middle of the galleries so that we would be able to mount central displays if we needed them. Electricians were called in and fixed conduits for the wiring across the under-floor space before the concrete was poured.

"Are you sure these are going to be OK?"

"Don't you worry, Dokotela, they will be alright on the night, isn't that what you say?"

The concrete was cast over the whole area and painters completed the walls. Outside, the royal wives from Lobamba constructed a Swazi village, erecting traditional beehive huts, complete with burnished cow dung on the floor. These traditional homesteads were fast disappearing from the rural scene, so it would be interesting for visitors to see how people had lived only ten or twenty years earlier, before corrugated iron and aggregate blocks had altered the landscape.

Christoph became really interested in the design and construction.

"You should landscape the gardens by building some large mounds of earth and putting granite boulders on top of them, to give the atmosphere of highveld scenery. Then you can add tree aloes and whatever around them. I'll do it for you." And he did.

Driving in through the front gates the whole approach began to look truly splendid. We planted rare, young, indigenous trees like yellowwoods around the edges. But my own greatest success was transplanting three two hundred year old kiaat trees donated by the Malkerns cannery. They were ploughing new pineapple fields and these old trees were in the way. When they were delivered, nearly all the limbs had been sawn off and the roots trimmed right back; they went into deep holes, with lots of fertilizer. They looked so dead I didn't think they would survive, but six months later strong buds began to push through the bark.

We had designed a veranda in front of the main exhibition hall with a roofed garden in front of that. Ralph had rescued some Umbuluzi cycads from the gorge which had been threatened by a new Swaziland railway siding and we planted them there.

The National Museum was nearly finished. Dick Klip had ordered the large letters that were going to spell out the title of the museum on the wall outside. These were two-foot high alloy letters with rods at the back to be placed slightly proud of the wall, giving a tasteful shadow effect.

I was by no means certain of the literacy levels of the workmen in charge of fixing letters. I had suffered nights of fitful dreams about the opening ceremony, pointing proudly at my creation and reading 'SWIZALAND

NITANAL MESUOMU'. I personally placed them in the right order and watched as the holes were drilled and the letters were cemented permanently into the plaster rendering. The perfectly shadowed titles read: SWAZILAND NATIONAL MUSEUM: *UMSAMO WESIVE*. Wonderful! It wasn't until after lunch, when the sun was oblique to the wall that I realised we had made a miscalculation. The letters were not all exactly in the same plane, so that their shadows appeared higgledy-piggledy across the wall, an effect that worsened as the afternoon wore on, but alas we couldn't change that now.

The carpet layers arrived and as the glue was being spread over the vast area of the concrete floor the electricians came to set the sockets in the ground. As they pushed the wires down the subterranean conduits another nightmare came true. An adenoidal youth called Mandla said:-

"It's the pipes! I can't get the wires through the pipes! They won't come out the other end! We're going to have to dig up the concrete!"

They brought in jack-hammers and began attacking the specially laid, never-to-be-disturbed concrete slab right in the middle of the new exhibition halls. The interior became carpeted not with quality brown cord, but with a thick layer of powder. It was the most depressing event of the project. But they found the break and the disturbed part of the concrete was re-laid, the carpet stuck down and the new exhibits hung. It was finished. The new main gallery showed a photographic history of Swaziland and the royal family through the ages, an appropriate exhibit for the king's jubilee, now only days away.

There was to be an official opening, covered by Swaziland TV. Dignitaries had been invited and the deputy Prime Minister, the Hon. Ben Nsibanze, was to perform the ceremony. The *Umsamo* was looking pristine. Two hours before the grand moment, two painters fell through the roof of the veranda and a road grader cut through the main water line; for a short while we had unscheduled fountains playing in the garden, but both mistakes were rapidly fixed.

The opening was a great success, with the exception of the fossil man case. As I described the incredible significance and globally-unifying implications of the exhibit, the Hon. Ben, whom I had forgotten was a lay preacher in Mbabane, looked at the casts of ancient skulls and said:-

"Dokotela, if you wish your family to be descended from these primitive creatures, that is your prerogative. But my family certainly are not. We are Methodists."

Human evolution notwithstanding, the Swaziland National Museum was a reality and proved to be one of the main attractions of the king's jubilee celebrations.

CHAPTER TWENTY

BEGINNINGS AND ENDINGS

The beginnings and endings of all human undertakings are untidy.
                                                      John Galsworthy

At the time of the jubilee, Sobhuza, by now more than eighty, had been on the throne for the past sixty years and was the oldest ruling monarch in the world. He had withstood the humiliation of colonial administration, the political upheavals of a world war and the indignity of treating with apartheid South Africa, yet he had successfully guided his tiny Swazi nation to its independence. He was one of the last unashamedly African rulers, but simple of dress and wise of counsel. Some years before, one of our teams had actually seen the 'Old Man', as he was affectionately known. We had been travelling in a Land Rover through a wild part of the lowveld bush near the Ngwavuma river in the Shiselweni district, the Place of Burning. It was an area of immense tradition and historic resonance, with imposing leadwood trees and tall hyperenia grass - old Africa. On a dusty track deep in the bush a dilapidated notice read 'No photographs,' which seemed incongruous in such a remote place.

"Hey, why the notice, Doc?" came the question from the back of the Land Rover.

As we passed a group of beehive huts I shouted that this was one of the king's 'royal residences.' They were thatched with local grass and no doubt floored with burnished clay from termite mounds. One of Sobhuza's many wives lived here and the king liked nothing better than to escape into the bush with his older consorts, away from the jangling cut and thrust of government at Lobamba.

At that moment around the corner of the bush track came two motor cyclists with blue lights flashing, followed by a Swazi police car with 'Royal Escort' on the roof, behind which came a black, somewhat dated limousine, then another police car and a two further motor cyclists. We pulled onto the grass at the side of the track and amid the swirling sand saw the Ngwenyama, Sobhuza II himself, sitting in the back of the limousine, a small, slight figure dressed in a single cloth drape. I raised both hands respectfully. He smiled regally at us and waved back and in a further eddy of dust the procession was gone.

Our group had come half way around the world and amazingly had just witnessed a traditional African potentate in his natural surroundings. No one spoke for several minutes, until someone suddenly asked,

"Say, David, was that an American car back there?"

"That," I said firmly, "was the world's oldest surviving monarch!"

"Well, I want you to know that whoever that old guy in the back was he was riding in probably the world's oldest surviving Cadillac limo!"

* * * *

The year after the jubilee, King Sobhuza died. He had been a good and a great ruler, a lion to his people, and a traditionalist to the end. I was reminded of his own words, that no one ever died a natural death; it is always by sorcery. We shall never know if he met his own end that way, but he was eighty three and had enjoyed a remarkable reign. The search for his successor began. It was to prove a most complicated affair which during the inter-regnum engendered a growing instability.

In 1966 Sobhuza had married one of the last of his one hundred and thirty or so wives, a young girl, Ntombi, of the Tfwala clan, a minor family among the Swazi hierarchy. Sobhuza, then sixty seven, said that he had had a dream in which his ancestors directed him to take a girl from that clan to build his 'house,' the emaDlamini royal dynasty. Ntombi fell pregnant and in 1968 gave birth to a son at the Raleigh Fitkin memorial hospital, Manzini, the very same one where Watty and I had the unnerving experience with the pregnant woman that night years before.

The late king had decided to call the child plain Nkhomotiphumakamlambo, but the princess charged with naming the baby, on her way hot-foot to the laying-in ward at Manzini, forgot this pretentious moniker and settled for the less demanding Sigidzankhundleni. That wasn't right either so he was named Tikhulutihlangene. Amazingly this infant had already got through over thirty consonants and twenty vowels and was still only a few hours old. But the matter didn't rest there. One of the king's senior wives renamed the new-born yet again, Makhosetive, and so he remained until the month the old king died, when the royal inner council agreed he should succeed his father, Sobhuza. From then on he was known simply as 'uMntfwana,' 'the Boy.'

Why was he chosen above the dozens of other princes, also the fruit of the late king's loins, born earlier and from more prominent queens? Some of them were by now in their sixties and could hardly be expected to hold the emaDlamini fort for very long; indeed, some had already died. Much bumping and boring did ensue but after protracted and fraught machinations 'the Boy' won by a short head, thanks no doubt to the fact that the Tfwala clan is insignificantly small; only five Tfwalas appear in the Swaziland telephone directory, even if you include all of them. Compare this, say, to the Nxumalo clan, which is very powerful and much more numerous; more than two hundred and sixty names are listed in the directory. Also, for class and local colour, Nxumalo has a soft click in it, N-click-umalo. So why not an Nxumalo to take the purple?

The first thought was that the new candidate should not yet have reached puberty, so there would be no unforeseen offspring to muddy the pools of Lobamba politics, already quite murky enough. It was also felt that the minor status of the Tfwala clan meant that having no strength or length to the Tfwala suit, 'the Boy' could be manipulated more easily by the powerful *bemdzabuko* chiefs, the core of the nation, and their all-controlling wives. He was considered a malleable compromise, hence he was the one eventually selected.

The aegis of Sobhuza and all the emaDlamini kings way back to Mswati I, Somhlolo, Ngwane and on into the mists of prehistory, would fall upon 'the Boy's' young shoulders. To scotch all the bitter infighting and the near-fatal confrontationalism which characterised the years immediately after his selection, it was decided to install the young man quickly before the emaDlamini household engaged in further internecine warfare. He would take the throne name Mswati III, after some of his illustrious predecessors. The coronation was arranged.

∗ ∗ ∗ ∗

"There is a rumour that the king's coronation may take place next Friday," I announced to an evening meeting of one of the last teams of our international volunteers. "Should it be so, we would take the day off."

One woman who had suddenly taken up journalism combatively declared; "What do you mean '*may* take place'? Don't you know when it will be? I have to leave for the U.S. on Sunday. What about my readers? This is awful, truly awful!"

She was from somewhere like Boiling Springs, Pennsylvania, and was hoping to write a free-lance piece about Swaziland for something like the *'Boiling Springs Bee.'* The coronation would be her first major foreign scoop.

I explained that not even the king-to-be knew exactly when he was to be crowned, or the venue. This was Swaziland, where Swazi lore and custom prevailed, the two bastions of cultural justification for the ever-growing armies of the corrupt and inefficient. It was all to do with omens, auspicious times, traditional healers, ritual emetics; all part of the long and tortuous heritage of the Swazi nation.

But it indeed turned out to be on Friday, with a public ceremony to be held in the national stadium the following day.

"So how do I get in? I need to see everything. The paper's readers need to know everything. Do you know exactly what is going to happen?"

I had to disappoint her again. The last 'coronation,' of the new king's late father, had taken place in 1921. A few elderly Swazis might have a distant memory of it, but no-one else. I would see what I could do.

Ikbal, the only Indian at that time allowed to open a shop in Swaziland, came to my aid. Incidentally, the Swazi government felt that if allowed to, Ikbal's brethren would take over every enterprise in the kingdom, as they had done everywhere else in southern Africa. Swaziland, being a totally multi-racial state, in distinction to the then apartheid republic to the west, allowed everyone across its borders without let or hindrance except Indians, who were normally only given a twenty four hour visa.

Ikbal had side-stepped this restriction and had the ear of princes high in the Swazi establishment. He got me unmarked tickets to the stadium, the diplomatic reception, the garden party, the royal cattle byre and anything else that pertained to this grand event. I filled in the name and a specious accreditation; 'Betty B. Boopenburger, Managing Editor, *Boiling Springs Bee*; affiliated with the *New York Times, Washington Post, Le Monde*, etc.'

His Majesty was crowned at Luzidzini, his new residence near Lobamba, spiritual heart of the nation. It took place at the large royal cattle byre, inside a smaller, even more royal cattle byre. This was a sacred event, witnessed by very few people, and conducted by the national priests of the Mkhatjwa clan. The world's press was also there in a separate part of the cattle byre, grateful that the rain held off otherwise they would have been up to their knees in it. Among the journalists was our own budding news hound, 'Betty B. Boopenburger' of the *Boiling Springs Bee*.

The ceremony ended with the famous announcement by Prince Mboni, advisor to the Ndlovukazi, the Great She Elephant, the king's mother:

"People of the Swazi nation, I am instructed by the elders to inform you that as you have been patiently waiting for the sun *(ilanga)* to rise, it has now risen!"

The following day, in the Somhlolo national stadium at Lobamba, the people were about to be shown their new monarch. This was the same stadium where earlier the same year that mega football team 'The Sidwashini Dribbling Wizards' had played Manchester United. The score reached telephone number proportions, giving rise to the headline in the *Times of Swaziland* by their ace soccer reporter Magasuthu 'Sports' Ngwenya: "Wizards destroyed by Man!"

Swazis in traditional dress had descended in their tens of thousands from their grass kraals in the faraway Komati valley, or trekked on foot from the torrid bushveld of Shiselweni and the distant crests of the Lubombo mountains adjacent to Mozambique. Everyone was packed in and waiting.

After a while the foreign dignitaries arrived. Maureen Reagan, daughter of then President Reagan, was the U.S. representative. Prince Michael of Kent represented the Queen of England. Also present, amazingly, was the poker-faced P. W. Botha, State President of the Suid Afrikaanse Republiek, one of the prime architects of the black homelands policy. Here too was Kenneth Kaunda of Zambia, self-appointed High Panjandrum of black politics in southern Africa. He was one of two leaders to make a speech, abstruse as ever, bombastic, prolix and utterly incomprehensible. Mozambique was represented by Samora Machel whose widow, ten years later, would marry Nelson Mandela.

Botswana, Lesotho and all the other African states sent representatives. They all arrived at the stadium in a special fleet of fifty identical black Mercedes 500's donated at great expense, to be sold afterwards no doubt at even greater profit. The saloons had no number plates, only the country name of the passenger - Zimbabwe, Germany and so on. A royal escort accompanied them from Mbabane, the administrative capital, down the dreaded Malagwane Hill, site of so many accidents. Whoops! There was an accident among the dignitaries. Inexperienced drivers had followed too close. France had collided with Botswana. No damage was done. The stadium was full to bursting, and the ceremony about to begin.

At noon, the newly installed Ngwenyama, King Mswati III, entered the stadium standing on the tail-gate of a white Land Rover bakkie and dressed in a leopard-skin loincloth, an impressive array of ostrich feathers on his head. The crowd went wild. He took his seat on the rostrum next to Maureen Reagan, dressed in a stylish hat and matching lime-green twin set. The nation was about to pay homage to its youthful king. He was eighteen years old.

First on were the Swazi maidens come to dance before their young Ngwenyama, paramount chief of the clans of the Swazi people. Bare-breasted and well-endowed, they wore low on their waists provocative silver-beaded fringed skirts the size of pelmets which barely cover their vital parts. From behind they exhibited a maidenly version of builder's bum. Around their ankles they had fastened seed pods filled with gravel and had fashioned multi-coloured bead pendants at their throats and tied bright, long woollen tassels from their waists and shoulders. They carried kitchen knives, flash lights, tiny leather shields or reeds in their hands and sporadically blowing whistles, they sang praises to the new king and stamped their feet so that the seed pods rattled. They were beautiful - large, small, stern, smiling - swaying in serried ranks, each one hopeful no doubt of being chosen to become the wife of the new royal household, for the king must marry with a daughter of each of the one hundred and thirty five or so clans of the Swazi Nation. Lucky king! The ageless granite hills surrounded the ceremony and bore silent witness to the continuing traditions of this proud people. In the background rose Sheba's Breasts, ascending in soft silhouette behind the stadium.

After what felt like hours, there was a break. The giggle of girls departed and the band of the Swaziland Defence Force (director of music, Len Clutterbuck MRCM), togged out in guards' black trews with red piping and wearing red tunics, struck up a traditional march, 'Stand up! Stand up for Jesus' in the key of C, for ease of playing. Their marching was faultless, if a trifle jazzy. Hip-swivelling is a way of life for Africans, even if it was to the solemn hymn tune 'Morning Light.'

Outside the stadium the food vendors and general hawkers were doing a roaring trade. Three-legged cauldrons were full of steaming ox offal. Irrigation-schemes-worth of traditional 'Warrior Beer' were being consumed from cardboard cartons marked 'Warrior Beer' with a picture of a Swazi ox-hide shield on the side. The drinkers were Swazi warriors in traditional dress carrying shields made from their piebald Nguni cattle. The warriors were gearing up for their afternoon performance with a little industrial-strength lubrication. The beer tasted and looked like a sort of fermented sour porridge. Remember Eugene from Yuba City, Sacramento?

In the stadium the pageant moved on. The *emakhosikati*, the royal wives and hundreds of other senior Swazi ladies danced barefooted before the king, wearing their traditional heavy, black, pleated skirts made from cow hide. Over their shoulders and to their hips they were bound with black and burgundy cotton wraps decorated with motifs of Swazi shields and assegais. They moved as one heaving mass of colour; each with her hair in a huge spherical glistening beehive, traditional mark of a married woman.

They swayed and sang traditional Swazi praise songs in a harsh yet harmonious descant and ululated with enthusiasm for their new young spiritual ruler, Mswati III. As they danced they were assuredly plotting the power struggle that would inevitably unfold between themselves, the *emakhosikati* of the late king, and the new generation of clan-derived wives. Swaziland being partly a matriarchal society, the clan elders machinated through their daughters chosen as the royal spouses.

The royal wives retired to conspire. The Swaziland Defence Force band reappeared. Another few minutes of hip-swivelling 'Standing up for Jesus' ensued before the Swazi warriors, buoyed up on a flood tide of beer, entered for the climax of the ceremony. They wore grey duiker-skin loin cloths and dark-patterned cotton skirts, under which one might occasionally glimpse a pair of Y fronts; they had fashioned bands over their exposed shoulders and bead adornments around their necks in multi-coloured panels of light blue triangles and white and black chevrons. Some had leopard-skin skirts handed down by their fathers - there were few if any leopards in Swaziland now. Others wore monkey-skin headgear and Timex watches.

They brandished knobkerries, a sort of Irish shillelagh cut from the branch and bole of a tree, with a long handle and a fearsome bulb at the end. Some sticks had a truck wheel-nut instead of the wooded bulb; think of the damage that could inflict. Traditional spears, the long-handled assegais, were banned in this ceremony. In similar previous events there had been too many accidents, especially after so much beer. They each carried a full length Swazi ox-hide shield, in black and white, brown and white, dappled, pure black, cream white, and formed into their impis, their age regiments and placed the king in their midst. Mswati had been educated in England, but nothing in the hallowed British traditions of his honeyed limestone school nestling among the Dorset Downs in mediaeval Sherborne would have prepared him for this ancient African extravaganza. Conspicuously, his headdress contained the iridescent metallic-red wing feathers of the g*walagwala* bird, the insignia of the emaDlamini, the clan from which all the kings of the nation must derive. The ceremony continued.

Joining the king and his warriors was the ever-corpulent and resplendently clad figure of the Zulu king Zwelithini. He and his party were decked out in leopard skin loin-cloths and headbands. Ironically, it was his own ancestor, Shaka Zulu, who had forced the Swazis to flee north of the Pongola River almost two hundred years before, probably harrying them with their own assegai blades. Now the two kings danced in solidarity until the sun angled down the sky and the ceremony stuttered to an end.

* * * *

With the king's coronation, a new order arose out of the ashes of the old, while our own time in Swaziland, in this blessed and beautiful country, was drawing inexorably to its close. The pageant of the new king's inauguration marked a watershed, culturally and politically, in the life of the kingdom, the Commission and of the Swaziland Archaeological Research Association. Ultimately, the king's beginning turned out to be our own ending. We had been working for more than ten years in Swaziland; our research had reached its natural conclusion and in any case we were running out of new ideas. Also, our own children would soon be in their teens, when they would need more permanent schooling in Britain. We all sensed that it was time to leave, to find new problems to solve and new regions to investigate. With heavy heart the team split up and we all went our separate ways.

But it was hard for me to stay away from Africa. It had become my addiction on that first magical, cacophonous, star-shining night, the night when Africa had first spoken to me. Swaziland was the means by which my craving had been satisfied. Africa had entered into my blood then; it courses through my veins still. It will always draw me back. Back to the highveld on that sun-filled day at Ngwenya where the air was like a draught of champagne and the chacma baboons tumbled helter-skelter down the ridge into the forest glade. Back to the bushveld, when at the end of the dry season the knobthorns flower, filling the air with their unmistakably sharp, intoxicating perfume. These are memories that will never leave me.

****

Come back with me now one last time, back to Sheba's Breasts shimmering in the sultry heat above Lobamba, back to the Swaziland National Museum nestling beneath the twin peaks of Lugogo. There are the ancient kiaats we planted over thirty years ago, boughs fully restored and thick with foliage, casting a deep shadow over the beehive huts of the Swazi homestead. Look! The yellowwoods have grown high and handsome, some of the finest I have seen. Can you see the tree aloes blooming over the granite boulders set among the well-kept lawns of the museum gardens? It is morning, and the shadows of the low-maintenance letters of the *Umsamo Wesive* appear to be straight against the wall. Ah, but no wonder; they've been reset flat so that no ghosted background offends the eye.

The museum has doubled in size. There, in one of the new galleries, stands an ancient black Cadillac sedan with rear fins and white-walled tyres, the one we saw all those years ago in the bush, with the oldest ruling monarch in the world reclining in the rear seat. Next to it another new gallery commemorates the wildlife of the kingdom and its game reserves, especially Mlawula. The whole atmosphere is poignantly impressive if static and lacking vitality.

There is a bus drawn up in the car park. I can make out the writing on the side. It's the Mhule One Way. This morning it has brought schoolchildren from the towns of southern Swaziland to Lobamba. They are only six and seven years old and have come to see the nearby memorial to the late king, Sobhuza II, who for them is already distant history. They have been visiting the parliament building around the corner and now they have come to the museum to learn about the rich heritage of the land of their birth. My God! Do you know how many visitors a year the Museum now gets? Over one thousand a day; that's a third of a million every year, mainly young Swazis! I'm speechless.

If we peer through the fronds we can see a few of them, dressed in their neat blue and black uniforms, with their pens poised and their paper questionnaires at the ready as their teacher tells them tales of old Swaziland, of a land long, long ago. Somewhere there are the Ngwempisi hand-axes made by some of Swaziland's first inhabitants, over a million years ago. There too is Jack's point, from deep in the trench at Sibebe, the rock-shelter at the top of that huge granite mountain. It was made maybe fifty thousand or more years in the past, but speaks to us still of the adventures of those highveld hunters.

Are the jewel-like scrapers of the Late Stone Age there and are Watty's Oo! Oo! Men, arms outstretched, still dancing across their stone stage? There are the shields of the old Swazi regiments, now disbanded, a memory of a world that has gone. The children are writing. They are learning about their inheritance. They compare notes and gaze with wide, liquid eyes at the cases and at the pictures.

Then, quietly, I sense the dull sound of voices around me. Can you hear them? That's the voice of Jules with Jane there, "Now come on, David! You know what that tree is!"

I can hear Nigel and Bruce, bickering about the TV presentation; "Oh for Christ's sake, Nigel, let him do it. He's a natural! It's a fabulous story!"

I can hear Larry, and Umbuluzi Two.

"That's a fine lawn out there, Doc. Mind you don't go out and piss on it by mistake!"

I can see in my mind's memory a picture of Barry, the Lemon Squeezy king, all excitement; and I can see the tears in Eugene's eyes, holding the past in his fingers. Remember him? "Man, this beer tastes like a rat's ass!" It probably still does, Gene.

I can hear Toko laughing softly.

"*Hawu! Bahbe*. What are you doing playing in the dust like that?"

There is Andrew.

And I can hear Colin's voice quite distinctly.

"Are you sure you're alright there director? Need a stiffener, do you? Another gin and toad, perhaps?"

As I watch, the school children gather round their teacher, and I hear Ralph's ghost speaking softly in my ear, "Come along Price Williams. No sense in getting over-emotional. Leave them to it. We've done our stuff. It's all theirs now. We'd better bugger off!"

And so we did.

AFTERWORD

To those who woke my sleeping memories, my thanks. I can't explain why, but some of the names I have used have changed slightly. In alphabetic order the authentic names are: Rosemary Andrade, Kenneth Alvin, Lawrence Barham, Larry Beleau, Jane Chard, Buster Culverwell, Bruce Dakowski, Sue Eagle, Mary Earnshaw, Sandra Eastwood, Shadrack Fakudze, Pat Forsyth Thompson, Ralph Girdwood, Andrew Goudie, Justin Grant, Cindy Henry, Roger Hooker, Caroline Jones, Simon Kumalo, Donna Lukshides, John Masson, James Matsebula, Rodney Maud, Margaret Maziya, Martin McKeowan, Juliet Prior, Richard Stone, Christopher Vickery, Andrew Watson and Betsy Woodwell.

A number of friends have read the manuscript and made suggestions, helpful or otherwise. Beside those immediately involved, they include Emma Bodossian, Angus & Small McLeod, Jemma Reynolds, Paul Slawson, Andrew Relph, Patrick Galbreath and last, but certainly by no means least, Wendy Vickery.

I must again thank the diligence of my brother John, 'Yr Hen Was,' for his superb job of editing. Any residual mistakes are my own.

But most of all, I have to thank my wonderful wife Sue who introduced me to Africa and who threw herself tirelessly into supporting the Swaziland Archaeological Research Association in word and deed, running the commissariat for a decade and particularly stimulating me to make a dramatic shift through the academic gears.

And to Alice and Daves, both of whom I know consider their time in Africa to be the pinnacle of their childhood, here is my version of what happened. I hope it brings back vivid memories.

Kew, January 2016

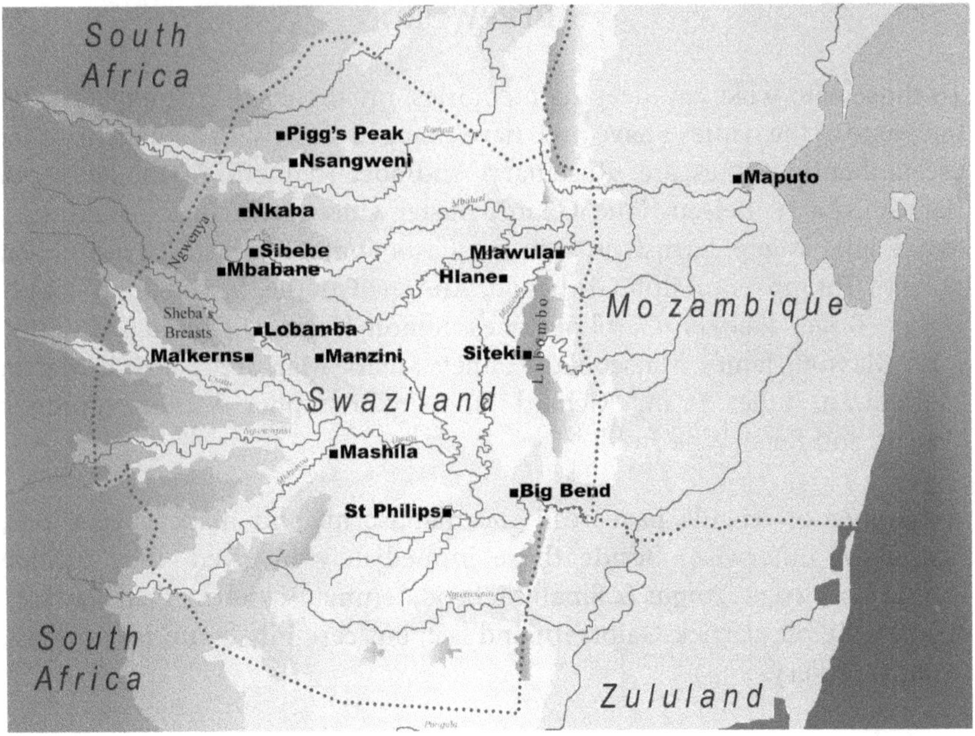

Map of Swaziland. (© DPW 2015)

Jack Conners and admirers at Sibebe with 'Jack's Point.'

The old farm and 'stoep' in Malkerns with two of the Expedition station wagons.

The Expedition Centre – 'The Long House.'

Infibulated male dancers and shamans at Insangweni Rock Shelter.

Above: Margaret 'Toko' Maziya.

Left: Watty making notes at a donga.

Andrew in Mashila Donga, Mhkondvo Valley.

Swaziland National Museum 2016.

'DPW' An oil collage by Betsy Woodwell mainly featuring Mlawula 1982.

www.ingramcontent.com/pod-product-compliance
Lightning Source LLC
Chambersburg PA
CBHW050524170426
43201CB00013B/2069